Dielectrics 2013

Journal of Physics: Conference Series Volume 472

Reading, United Kingdom
10-12 April 2013

ISBN: 978-1-62993-450-1
ISSN: 1742-6588

Printed from e-media with permission by:

Curran Associates, Inc.
57 Morehouse Lane
Red Hook, NY 12571

Printed by Curran Associates, Inc. (2014)

For permission requests, please contact the Institute of Physics at the address below.

Institute of Physics
Dirac House, Temple Back
Bristol BS1 6BE UK

Phone: 44 1 17 929 7481
Fax: 44 1 17 920 0979

techtracking@iop.org

Additional copies of this publication are available from:

Curran Associates, Inc.
57 Morehouse Lane
Red Hook, NY 12571 USA
Phone: 845-758-0400
Fax: 845-758-2634
Email: curran@proceedings.com
Web: www.proceedings.com

TABLE OF CONTENTS

PREFACE

This volume records the 42nd Dielectrics Group Proceedings of the Dielectrics Conference that took place at the University of Reading UK from 10–12 April 2013. The meeting is part of the biennial Dielectrics series of the Dielectrics Group, and formerly Dielectrics Society, and is organised by the Institute of Physics.

The conference proceedings showcase some of the diversity and activity of the Dielectrics community worldwide, and bring together contributions from academics and industrial researchers with a diverse background and experiences from the Physics, Chemistry and Engineering communities. It is interesting to note some continuing themes such as Insulation/HV Materials, Dielectric Spectroscopy, Dielectric Measurement Techniques and Ferroelectric materials have a growing importance across a range of technologically important areas from the Energy sector to Materials research, Semiconductor and Electronics industries, and Metrology.

We would like to thank all of our colleagues and friends in the Dielectrics community who have supported this event by contributing manuscripts and participating in the event. The conference has provided excellent networking opportunities for all delegates. Our thanks go also to our theme chairs: Dr Stephen Dodd (University of Leicester) on Insulation/HV Materials, Professor Darryl Almond (University of Bath) on Dielectric Spectroscopy, Dr John Blackburn (NPL) on Dielectric Measurement Techniques and Professor Anthony R West (University of Sheffield) on Ferroelectric Materials. We would also like to thank the other members of the Technical Programme Committee for their support, and refereeing the submitted manuscripts. Our community would also like to wish a full recovery to our plenary speaker Prof John Fothergill (City University London) who was unexpectedly unable to give his talk as well as thank Professor Alun Vaughan for stepping in and giving an excellent plenary lecture in his place at such very short notice.

We are also pleased to express our thanks to the Conference Department of the Institute of Physics for their invaluable support in organizing this event. We are especially grateful to Dawn Stewart for her responsive and day-to-day handling of this conference, as well as Claire Garland for help in planning and managing this international event. We would also like to thank Dr Steve Welch, Director at ESP Central Ltd, representing the interest of the Electronics, Sensors, Photonics Knowledge Transfer Network, as well as Paul Naylor and Susan Matos for their contribution towards the KTN session of the Conference. Finally, we would like to thank Solartron Analytical, Ametek and Princeton Applied Research for demonstrating precision electrochemical impedance spectroscopy measurement techniques at the conference.

We hope that the wider Dielectrics community will find these proceedings of interest and will use them as reference text in their future work.

Programme committee

R Pethig, University of Edinburgh
J Blackburn, National Physical Laboratory
J Swingler, Heriot Watt University
S Hadjiloucas, University of Reading
A West, University of Sheffiled
M Hughes, University of Surrey
S Dodd, University of Leicester
D Almond, University of Bath
M Cain, National Physical Laboratory
D J Swaffield, University of Southampton
N Green, University of Southampton
A Vaughan, University of Southampton

Sillas Hadjiloucas and John Blackburn (Summer 2013)

Peer review statement

All papers published in this volume of *Journal of Physics: Conference Series* have been peer reviewed through processes administered by the proceedings Editors. Reviews were conducted by expert referees to the professional and scientific standards expected of a proceedings journal published by IOP Publishing.

Dielectrics 2013 IOP Publishing
Journal of Physics: Conference Series **472** (2013) 012001 doi:10.1088/1742-6596/472/1/012001

In Praise of Simple Models

G van der Zwan

Department of Exact Sciences, VU University, de Boelelaan 1083, 1081 HV, Amsterdam, The Netherlands.

E-mail: zwan@few.vu.nl

Abstract. Simple models, in particular those based on Onsager's point dipole in a spherical cavity embedded in a polarizable medium, played an important role in understanding equilibrium and dynamical interaction between solvent polarization and classical reaction kinetics. After a brief review of some crucial results, and an application to the interaction of fluctuating polarization field on a classical dipolar oscillator, I investigate the applicability of a similar simple model to quantum–classical interactions. Interesting differences with a fully classical system occur in equilibrium situations, and some fundamental problems with commonly used dynamical approaches are indicated.

1. The Onsager Model for an Oscillating Dipole.

Onsager's model for a dipole in a spherical cavity[1], although originally developed to better understand the relation between molecular dipole moment and macroscopic dielectric constant[1], also played an important role in understanding the role of solvents in chemical reaction kinetics. Both the theory of dielectrics and that of reaction rate constant made considerable initial steps in the 1930's, and continued developing somewhat in parallel until the early 1980's. An extremely brief history of the theory of reaction rates lets it start in the 1930's with transition state theory[3], followed by the realization that solvent dynamics in the barrier region can play a crucial role by Kramers in 1941[4], and the understanding of dynamical interaction between solvent and reaction rates in the late 1970's[5]. Marcus' theory for electron transfer rates fits in this picture[6], since it emphasizes the fundamental role of polarization in creating the barrier for an electron transfer reaction taking place in a polarizable medium.

The model is shown in Fig. 1. A point dipole is located at the center of a spherical cavity of radius a. For our purposes it is sufficient to take vacuum inside the cavity, whereas the dielectric can be described by the electric field E_o, the displacement D_o, and the polarization P_o. Since, apart from the point dipole, there is vacuum inside the cavity, we only have to consider the inside electrical field E_i. At the boundary these fields are related by boundary conditions

$$\vec{n} \cdot \vec{E}_\mathrm{i} = \vec{n} \cdot \vec{D}_\mathrm{o} \quad \text{and} \quad \vec{n} \times \vec{E}_\mathrm{i} = \vec{n} \times \vec{E}_\mathrm{o} \tag{1}$$

[1] However, Onsager's remark that Debye neglected to disregard the reaction field in his calculation is incorrect. Debye used the internal field calculated by Lorentz, who ignored the change in a dielectric due to a cavity. Onsager calculated the internal field by taking the boundary conditions at the cavity surface into account. It is the change in internal field, and not the reaction field that gives the Onsager relation between dielectric constant and molecular dipole moment. An interesting derivation of the Lorentz field can be found in Feynman [2]. The Lorentz field does not contain the reaction field.

Relations between the various fields, and the frequency dependent dielectric constant $\epsilon(\omega)$ are shown in the figure. We will denote the static dielectric constant by $\epsilon_s = \epsilon(0)$.

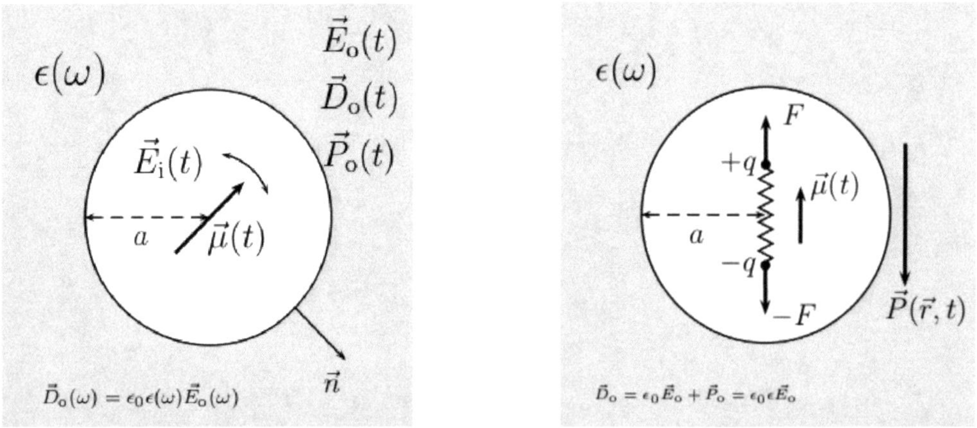

Figure 1. Two dynamic extensions of the Onsager cavity model. On the left hand side a model used by Nee and Zwanzig[7] to describe dielectric friction on a rotating dipole. A dipole with time dependent orientation polarizes the environment. Since the outside polarization needs time to adjust to the new situation the reaction field lags the dipole, causing friction. On the right hand side a similar model where the dipole does not rotate, but can change its magnitude. Two charges q and $-q$ are connected by a spring, forming a dipolar oscillator. The dipole moment causes an electric field that polarizes the environment, and this polarization causes the reaction field, which has two effects on the dipolar oscillator: its frequency is lowered, and it experiences a friction.

The internal field $\vec{E}_i$ consists of two parts: the vacuum field due to the charge distribution at the center, and a *reaction field* $\vec{E}_R$ due to the outside polarization. When the central charge distribution depends on time, all fields do likewise. The simplest form of dynamics of the polarization field is the Debye model where the polarization can adjust locally — by rotation of the solvent dipoles — to a new situation. This can be described by a Langevin equation:

$$\frac{d\vec{P}_{\rm o}(\vec{r},t)}{dt} = \frac{\epsilon_0\chi\vec{E}_{\rm o}(\vec{r},t) - \vec{P}_{\rm o}(\vec{r},t)}{\tau_D} \tag{2}$$

where $\chi = \epsilon_s - 1$, and τ_D is the Debye relaxation time, related to the rotational diffusion time of the molecules comprising the medium. This expression leads to the following frequency dependent dielectric constant:

$$\epsilon(\omega) = \frac{\epsilon_s - i\omega\tau_D}{1 - i\omega\tau_D} \tag{3}$$

Solving the boundary value problem gives the following expression for the reaction field:

$$\vec{E}_R(\omega) = \frac{1}{4\pi\epsilon_0 a^3}\frac{2(\epsilon(\omega)-1)}{2\epsilon(\omega)+1}\vec{\mu}(\omega) \equiv A(\omega)\vec{\mu}(\omega) \tag{4}$$

so that, at time t the energy $U(t)$ of a dipole in its own reaction field can be written as

$$U(t) = -\vec{\mu}(t)\cdot\vec{E}_R(t) \tag{5}$$

The time dependence of the reaction field is ultimately related to the dynamic equation satisfied by the polarization. Eq. (2) is only the simplest possible expression. More elaborate models have been considered within the context of reaction dynamics, such as the inclusion of inertial terms[8], or the possibility of non–local adjustment due to polarization diffusion[9]. These are all shown to lead to more or less complicated expressions for the function $A(\omega)$. The reaction field can be quite large. For a cavity of radius 0.5 nm, and a 5 D dipole (one electron charge displaced over 1 Å), it is already 1200 cm^{-1} in equilibrium.

Next we consider the effect of the reaction field on a dipolar oscillator, depicted on the right hand side of Fig. 1. In the presence of a polarizable medium outside the cavity the oscillator satisfies the usual equation of motion:

$$m\frac{d^2x(t)}{dt^2} = -m\omega_0^2 x(t) + F(t) \tag{6}$$

where m is its reduced mass, and ω_0 its frequency. $F(t)$ is the force acting on the dipole due to its own reaction field. An analysis closely resembling that of Nee and Zwanzig[7] leads to the following equation of motion for the extension $x(t)$ of the oscillator

$$m\frac{d^2x(t)}{dt^2} = -m\omega_0^2 x(t) + \mu^2 \int_{-\infty}^{t} d\tau\, A(t-\tau)x(\tau) \tag{7}$$

This can be written in a more interesting way by first considering the Fourier transformed equation:

$$-m\omega^2 x(\omega) = -m\omega_0^2 x(\omega) + \mu^2 A(\omega)x(\omega) = -m\omega_0^2 x(\omega) + \mu^2 \frac{A(\omega)-A_s}{-i\omega}(-i\omega x(\omega)) + A_s x(\omega) \tag{8}$$

followed by inverse Fourier transformation, which leads to:

$$m\frac{d^2x(t)}{dt^2} = -(m\omega_0^2 - \mu^2 A_s)x(t) - \int_0^t d\tau\, \zeta_D(t-\tau)\frac{dx(\tau)}{d\tau} \tag{9}$$

In these equations $A_s = A(\omega = 0)$, and $\zeta_D(t)$ is a friction kernel, defined by its Fourier transform

$$\zeta_D(\omega) = \mu^2 \frac{A(\omega) - A_s}{-i\omega} \tag{10}$$

Eq. (9) shows that the external polarization has two basic effects: it modifies the vibrational frequency, and it causes a friction on the oscillator, which will eventually bring it to rest. The modification of the frequency is usually not very large. In the case of an OH vibrational mode in a highly polarizable environment it can be a few tens of cm^{-1}, but OH modes in those circumstances are usually very broad peaks to begin with. Part of that linewidth could be due to polarizable solvent effects. For other vibrations the effect is usually much smaller, a few cm^{-1} at most.

More interesting is considering this equation from a fundamental point of view. The oscillator at the center of the cavity is itself undamped, it will oscillate forever if there is no interaction with the external field. The polarization outside the cavity is damped — in fact for the case I considered here even overdamped — and only because of the coupling the oscillator becomes damped as well. In fact a much stronger statement can be made. If we add random polarization fields to Eq. (2), and assume a fluctuation dissipation theorem to be valid[10] it is a straightforward, albeit lengthy, exercise to prove that Eq. (9) is a suitable generalized Langevin equation. Which means that it also contains a random force term for which the fluctuation dissipation theorem holds. This has the strong implication that the distribution of x–values decays to the correct equilibrium distribution. It can also be shown that A_s is in fact the potential of mean force on the oscillator resulting from the equilibrium fluctuations of the outside polarization field.

Although calculations for systems based on this approach are simple, they contributed a lot to our understanding of processes occurring in solvents[8, 11, 12]. Many analytical results can be derived which help in understanding the various regimes in solvent reaction dynamics[13]. One serious drawback, however, is that the model is completely classical, whereas for many processes the quantum nature of the system can not be negelected. One would expect, or could hope, that a similar approach to mixed classical quantum systems — using simple but analyzable systems — would also help understanding classical–quantum interaction.

2. Equilibrium Interaction between Quantum Systems and Classical Fields.

Interaction between classical and quantum systems is usually described in the following way[14, 15]: the quantum system 'feels' a classical electric field due to polarization, and the classical system 'sees' the expectation value of some operator of the quantum system. For the latter usually the charge density is used. This is not according to the basic principles of quantum mechanics, which states that an observation leads to finding an eigenvalue of such an operator, upon which the quantum system collapses to an eigenstate. Unfortunately there are few papers in which this approach is used for simple models like the one described in the previous section, even though much could be learned from that.

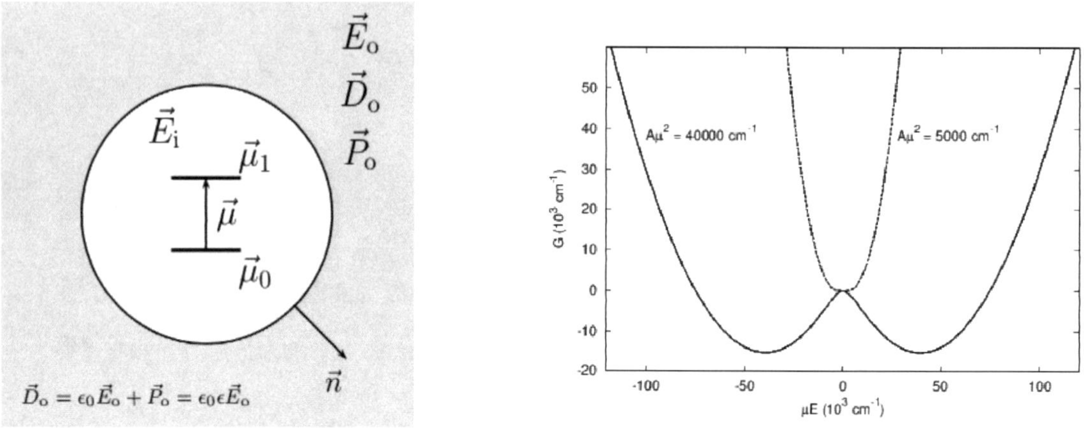

Figure 2. A two–level system (2LS) in a cavity. The properties of the cavity and the polarizable medium are the same as for the classical case, but now we have a quantum system at the center. This quantum system can have a ground– and excited state dipole moment, as well as a transition moment $\vec{\mu}$. The transition moment is responsible for the polarization. On the right hand side free energy curves for various values of the interaction strength A are drawn, for the case $\vec{\mu}_0 = \vec{\mu}_1 = 0$. For large enough values of A the free energy has minima at finite values of the ground state dipole moment. Om the right hand side free energy curves (see below Eq. (15)) for values below and above the splitting threshold. The ground state dipole moment at the minimum is 0.9 D. Input values were: ϵ = 12500 cm^{-1}, μ =10 D.

The set–up is explained in Fig. 2. Mathematically it can be described by the Schrödinger equation

$$\mathcal{H}\,|\Psi\rangle = \lambda\,|\Psi\rangle \quad \text{with} \quad \mathcal{H} = \mathcal{H}_0 - \hat{\mu}E \tag{11}$$

where $\mathcal{H}_0$ is the Hamiltonian of the unperturbed 2LS, $\hat{\mu}$ is the dipole operator, and E is the electric field acting on the 2LS. This electric field should be a reaction field, which means that it is caused by the presence of a dipole in the cavity. In an equilibrium situation this means that

$$E = A_s \langle \hat{\mu} \rangle \tag{12}$$

where A_s is the static reaction field factor introduced in the previous section. The two equations (11) and (12) need to be solved simultaneously. For the purpose of simplicity I take the simplest situation where there are no ground and excited state dipole moments, just a transition moment. The polarizability of such a quantum system is $\alpha = 2\mu^2/\Delta$ where μ is the magnitude of the transition dipole moment and Δ the energy difference between ground and excited state. Both equations can be derived from a single Hamiltonian

$$\mathcal{H} = \mathcal{H}_0 - \hat{\mu}E + \frac{E^2}{2A_s} \tag{13}$$

where a variational principle for finding minima leads directly to Eqs. (11) and (12).

Using a representation of $|\Psi\rangle$ in terms of the original wavefunctions $|0\rangle$ and $|1\rangle$:

$$|\Psi\rangle = \frac{t}{\sqrt{t^2+1}}|0\rangle + \frac{te^{i\phi}}{\sqrt{1+t^2}}|1\rangle \tag{14}$$

where t can be chosen real. Finding the eigenstates of Eq. (13) is equivalent to finding the extrema of the function

$$G = \frac{\Delta t^2 - 2\mu E t}{t^2+1} + \frac{E^2}{2A_s} \tag{15}$$

both with respect to E and t. This leads to solutions for t given by

$$t = 0 \quad \text{and} \quad t^2 = \frac{2A_s\mu^2 + \Delta}{2A_s\mu^2 - \Delta} \tag{16}$$

which shows that additional real solutions are possible when $2A_s\mu^2 > \Delta$. Those solutions give rise to a finite ground state dipole moment:

$$\mu_g = \pm\mu\sqrt{1 - \frac{\Delta}{2A_s\mu^2}} \tag{17}$$

This implies that the quantum system can acquire a ground state dipole moment if the interaction between the 2LS and polarizable solvent is large enough. This is not possible classically: the only solution of $\mu = \alpha E$ together with $E = A_s\mu$ is $\mu = E = 0$. In Fig. 2 on the right I show the free energy curves based on minimizing G with respect to t. For small values of A_s there is only a minimum at $\mu_g = 0$, but for larger values of A_s minima occur at the finite ground state dipole moments given by Eq. (17). For those values the higher energy of the ground state $|g\rangle$ is offset by the lower energy due to interaction with the environment $-\mu_g^2 A_s$.

As far as I know, no theory of dielectrics was formulated on the basis of this model system.

3. Dynamics of Interaction between Quantum Systens and Classical Fields.

Although the description of the interaction between classical and quantum system does not lead to obvious problems or inconsistencies, the same is not true for its dynamical extension. As indicated in Sec. 1 classical interaction leads to a consistent picture for the dynamics of the subsystem: it experiences a potential of mean force due to equilibrium fluctuations, and friction due to solvent relaxation. The interaction, as used for instance in QM/MM approaches to mixed dynamics, is similar to the one described in the previous section. The quantum system 'feels' the electric field due to the classical environment, whereas the environment experiences the expectation value of a quantum operator it is coupled to. Since the final equilibrium state is a mixed quantum state, it is mandatory to use the density operator formalism for this description[16], so we are led to the following equation for the density operator of the 2LS:

$$\frac{\partial \rho}{\partial t} = -\frac{i}{\hbar}[\mathcal{H}_0, \rho] - \frac{i}{\hbar}[\hat{\mu}, \rho]E(t) \tag{18}$$

In this case I consider the electric field $E(t)$, due to the outside polarization to be overdamped, so that it satisfies an equation similar to Eq. (2):

$$\frac{dE}{dt} = -\frac{E - A\langle\hat{\mu}\rangle}{\tau_D} \tag{19}$$

where $\langle\hat{\mu}\rangle = \mathrm{Tr}[\hat{\mu}\rho]$. This latter condition and the second term in Eq. (**??** makes the equations nonlinear in a rather complicated way. As in the classical situation, the equation of motion for the density operator is undamped, and it can be hoped that coupling to the damped equation of motion for the E–field would lead to decay of the density matrix to the ground state, or even to the equilibrium state if random fields with the correct properties are added to Eq. (19). This does not happen. Here just two numerically calculated traces for different initial conditions are presented. These indicate that none of the expected behavior occurs, and in fact show behavior seen in many non–linear systems.

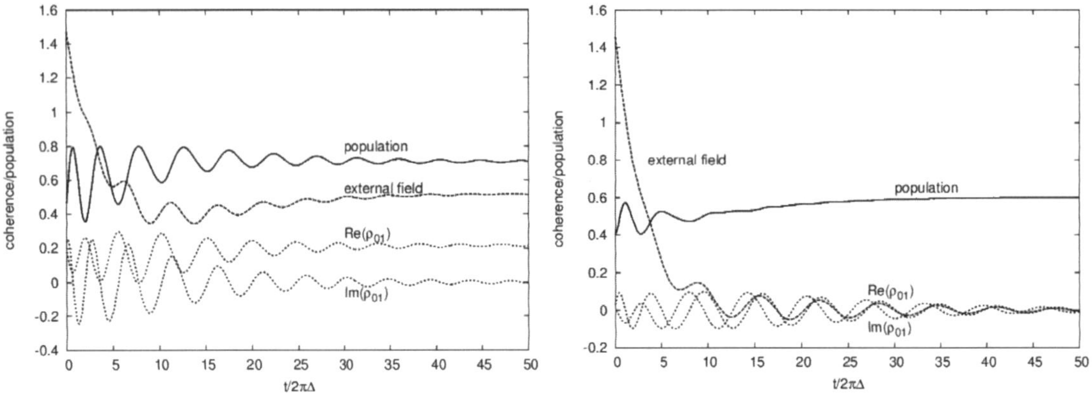

Figure 3. Two simulations of Eqs. (18) and (19) above the bifurcation threshold (*cf.* Fig. 2), with equal values of coupling– and dissipative constants, but with different initial conditions for the coherence of the density matrix. On the left hand side the coherence starts out at a finite value, and the ground state dipole moment (propoertional to the real part of the coherence ρ_{01}) and external field go to finite values. On the right hand side all initial conditions are the same, except that we start the coherence at 0. In that case both external field and expectation value of the dipole operator go to zero, which according to the results of the previous section should be a maximum in the free energy.

The behavior of the solutions is sensitively dependent on the initial conditions, which is often the case for non–linear equations[17], and also on parameter values. A full analysis of these, and similar, equations will be submitted elsewhere. They show in general complicated behavior, completely at variance with the similar simple classical models referred to in the first section. Since the methodology presented here is used in many QM/MM papers, where many other issues complicate the results — in particular the way classical and quantum systems are separated, and the pesence of many simultaneously coupled modes of motion — so that the problems found in these simple models are obscured.

The equations of motion described above can be derived from a mixed classical–quantum mechanical Liouville equation[16], with the classical part extended to a Fokker–Planck equation[10] in the high friction limit[4]. It was pointed out that such a mixed description leads to fundamental problems and inconsistensies[18]. The example given here shows that these inconsistencies have unpredictable consequenties for the solutions. The fundamental problem lies in the fact that classical phase space and quantum mechanical Hilbert spaces cannot be embedded in a single mathematical structure in which both can coexist, and where proper limits can be taken.

4. Concluding remarks.

Simple systems, in particular the ones based on Onsager's dipole in a cavity model, played an important role in understanding the influence of solvent dynamics in solution reactions. The advantage of such a model is clear: in many cases analytical results are possible and fundamental properties can be derived that give fundamental insight in the behavior of systems embedded in a polarizable solvent. Unfortunately these results are restricted to classical interactions. No systematic study appears to have been made for mixed classical–quantum interaction for similar simple systems, and the results derived in this paper point to numerous fundamental problems with this approach. These problems are caused by an imperfect understanding of how a classical system interacts with a quantum system, and so far our simple models give a clear indication of limitations of common approaches used in quantum chemistry and QM/MM methodology.

References

[1] Onsager L 1936 *J. Am. Chem. Soc.* **58** 1486–1493
[2] Feynman R, Leighton R and Sands M 1964 *The Feynman Lectures on Physics. Volume II.* (Addison–Wesley)
[3] Eyring H 1938 *Trans. Faraday Soc.* **34** 41–48
[4] Kramers H 1941 *Physica* **7** 284–304
[5] Grote R F, van der Zwan G and Hynes J 1984 *J. Phys. Chem.* **88** 4676–4684
[6] Marcus R 1956 *J. Chem. Phys.* **24** 966–978
[7] Nee T W and Zwanzig R 1970 *J. Chem. Phys.* **52** 6353–6363
[8] van der Zwan G and Hynes J 1985 *J. Phys. Chem.* **89** 4181–4188
[9] van der Zwan G and Hynes J 1983 *Physica* **121A** 227–252
[10] van Kampen N 1981 *Stochastic Processes in Physics and Chemistry* (Amsterdam: North-Holland)
[11] van der Zwan G and Hynes J 1982 *J. Chem. Phys.* **76** 2993–3001
[12] van der Zwan G and Hynes J 1983 *J. Chem. Phys.* **78** 4174–4185
[13] van der Zwan G and Hynes J 1984 *Chem. Phys.* **90** 21–35
[14] McCreery J, Christoffersen R and Hall G 1976 *J. Am. Chem. Soc.* **98** 7191–7197
[15] Miertuš S, Scrocco E and Tomasi J 1981 *Chem. Phys.* **55** 117–129
[16] Mukamel S 1995 *Principles of Nonlinear Optical Spectroscopy* (New York: Oxford University Press)
[17] Bar-Yam Y 1997 *Dynamics of Complex Systems* (Boulder CO: Westview Press)
[18] Caro J and Salcedo L 1999 *Phys. Rev. A* **60** 842–852

Finite Element Analysis of the Vertical Levitation Force in an Electrostatic MEMS Comb Drive Actuator

J Wooldridge, J Blackburn, A Muniz-Piniella, M Stewart, T A V Shean, P M Weaver and M G Cain

National Physical Laboratory, Hampton Road, Teddington, TW11 0LW, UK

E-mail: `jenny.wooldridge@npl.co.uk`

Abstract. A vertical levitation electrostatic comb drive actuator was manufactured for the purpose of measuring piezoelectric coefficients in small-scale materials and devices. Previous modelling work on comb drive levitation has focussed on control of the levitation in standard poly-silicon devices in order to minimize effects on lateral modes of operation required for the accelerometer and gyroscope applications. The actuator developed in this study was manufactured using a 20 μm electroplated Ni process with a 25 μm trench created beneath the released structure through chemical wet etching. A finite element analysis using ZINC was used to model electrostatic potential around a cross section of one static and one movable electrode, from which the net levitation force per unit electrode was calculated. The model was first verified using the electrode geometry from previously studied systems, and then used to study the variation of force as a function of decreasing substrate-electrode distance. With the top electrode surfaces collinear the calculated force density is 0.00651 $\epsilon_0 V_M^2 \mu m^{-1}$, equivalent to a total force for the device of 36.4 μN at an applied voltage of V_M=100 V, just 16% larger than the observed value. The measured increase in force with distance was smaller than predicted with the FEA, due to the geometry of the device in which the electrodes at the anchored ends of the supporting spring structure displace by a smaller amount than those at the centre.

1. Introduction

PiezoMEMS devices incorporate the use of piezoelectric films [1, 2] into Micro-Electro-Mechanical Systems for novel, miniaturised components such as resonators [3] and filters for wireless applications [4], switches in RF applications [5], inertial sensors [6] and energy harvesting devices [7]. Successful commercialisation of such applications requires accurate metrology for evaluating the properties of the piezoelectric films. Whilst there are several optical methods of quantifying the strain response of micro-scale piezoelectric materials to an applied voltage, the measured displacement is convoluted with the substrate clamping effects of the underlying silicon layers or substrate. The measurement piezoelectric response will therefore contain derived quantities from the mechanical properties of the silicon, which are not always easily determined. As such, a direct measurement of the piezoelectric material is preferable.

To this end, a MEMS piezoelectric measurement system was designed, built and tested for the evaluation of piezoelectric coefficients in small-scale systems. The system consists of an electrostatic comb drive actuator, which imparts a known force to the sample under test, and isolated electrical routing to an off-chip, ultra-low charge measurement circuit, to calculate the piezoelectrically induced charge. Full details of the MEMS design and operation can be found in reference [8]. A simulation of the free displacement of the vertical levitation actuator is shown in figure 1. A central plate is under-etched and suspended above the substrate by means of eight serpentine beams. 'Finger' electrodes of width 10 μm and lengths between 150-400 μm were attached to the supporting beams, and interdigitated with equivalent electrodes pointing towards the centre of the device, attached to the substrate.

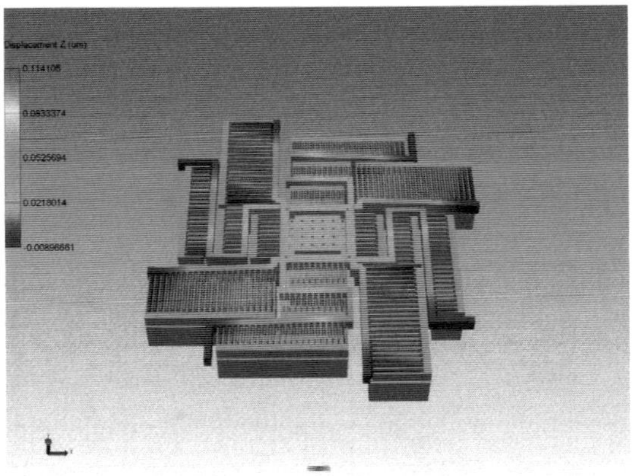

Figure 1: (Colour online) Simulation of the free displacement of the actuator with an applied voltage of $V_M = 50$ V. The colour map refers to the levitation height of the suspended MEMS structure.

Vertical levitation occurs within a comb drive actuator because of an asymmetry in the geometry and electrical bias within the structure. Figure 2 shows a cross section of a set of comb fingers: the stationary electrodes are biased to a voltage of V_M, and the movable electrode is electrically grounded along with the silicon substrate. A vertical force on the rotors is established in the direction away from the substrate, due to the interaction between surface charges on the electrodes and the surrounding potential, which aim to pull the comb finger upwards to a height at which the voltage seen by the top and bottom surfaces is equal. In reality this equilibrium position is dictated by the interplay between the electrostatic force, the restorative spring force from the flexural supporting beams of the device which anchor the comb fingers to the substrate, and the force due to gravity on the suspended structure.

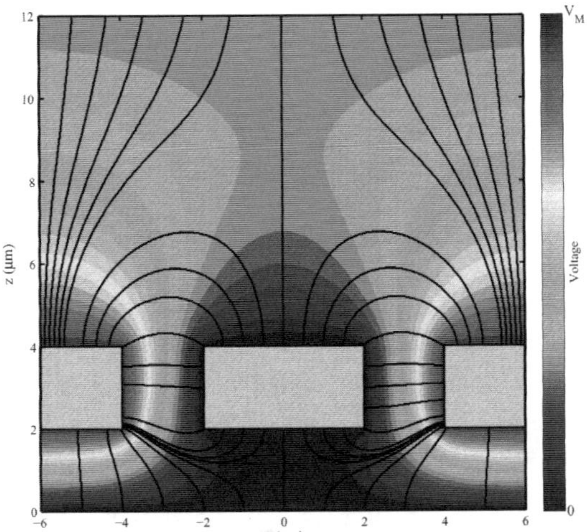

Figure 2: (Colour online) Plot of the electric potential (colour contours), and the electric field orientation (thick black lines) expected when the stationary electrodes (outer two grey rectangles) are set at V_M, and the movable electrode (middle grey rectangle), the upper edge of the simulation (not shown in diagram) and substrate at z = 0 are grounded.

In the piezoelectric measurement, the sample is clamped directly above the MEMS device, and the fully blocked force is transferred to the piezoelectric. Care must be taken not to significantly depress the central pad of the MEMS device with the contact to the sample, as it will have an effect on the size of the applied force. Since all previous investigations of comb drive levitation [9, 10, 11, 12, 13, 14, 15, 16] have looked at the variation in force with positive displacement from the neutral position (because the force is always away from the substrate, regardless of the polarity of V_M), it was necessary to simulate the effect of movements towards the substrate for the full characterisation of the device described above.

2. Experimental Details

A finite element package developed in house at NPL [17] was used to compute the levitation force on the movable electrodes. The problem to be solved in this case is to determine the electrical potential field through the cross section of the comb fingers, and from this to derive the electric and displacement fields. Following the methodology of Chyuan et al. [13], we evaluate the electric field normal to the movable comb finger surfaces (E_n, $n = 1, 2$) and calculate the net levitation force per unit length of electrode as:

$$F_n = \int_{\boldsymbol{B}} -\frac{\epsilon_0 E_n^2}{2} d\boldsymbol{B} \tag{1}$$

where $\boldsymbol{B}$ is the line across the horizontal edge of the conductor. The total force is simply the difference in the path integrals between the top and bottom surfaces.

3. Results

To check the validity of the model, the code was first run with the geometry evaluated by Chyuan et al. [13], and experimentally tested by Tang et al. [18], where electrodes of width 4 μm and thickness 2 μm are laterally separated by a distance of 2 μm, with a gap to the substrate of also 2 μm. For the finite element model, a 2D mesh was created 40 μm in height and 6 μm in width. Nodes were created every 0.1 μm in he lower half of the mesh, with the z step increasing to 0.5 μm in the top half, far away from the electrodes where the electric field is expected to be small. In the simulation the following Dirichlet boundary conditions were applied. All nodes within the stationary electrodes were set to 1 V. The nodes within the movable electrodes, and on the top and bottom edges of the mesh (the air furthest away from the comb fingers, and the grounded substrate respectively) were set to 0V. The voltage at all other nodes was allowed to vary over 2×10^5 iterations, obeying the relation:

$$-\nabla \cdot (\epsilon\epsilon_0 \nabla V) = 0 \tag{2}$$

Examination of the initial and final energies of the system indicated that a solution had been found to good accuracy.

The electric field across the top of the bottom surfaces of the movable electrode was calculated by the gradient of the electric potential in the z direction, and is shown in panel A of figure 3. The results closely match those presented by Chyuan et al. [13] who evaluated the electric field using a boundary element method (BEM). These data were used to evaluate the net vertical force per unit length of electrode as described in equation 1, found to be 0.1143 $\epsilon_0 V_M^2/\mu m$. The simulation was then repeated with the central electrode translated upwards in z. As can be seen in panel B of figure 3 the levitation force passes through zero at a point where the electric field intensity around the top and bottom of the device is equal. Moving the electrode above this reverses the sign of the force, pulling the suspended plate back towards the substrate. The equilibrium point was calculated to be 1.12 μm, in comparison with a value of 1.19 μm predicted by Chyuan et al. and the experimentally verified position of 1.22 μm by Tang et al.

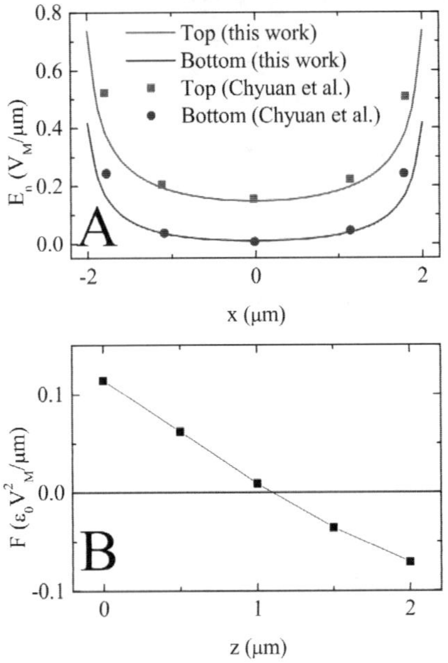

Figure 3: (Colour online) Panel A: The magnitude of the electric field along z along the top and bottom surfaces of the conductor (solid lines). The data reproduced from Chyuan et al is plotted using squares and circles for the same quantities. Panel B: Net vertical force density acting on the movable electrode when the comb finger has been displaced vertically by z.

The FEA input files were then adapted to describe the size of the devices manufactured in this study, with 10 μm wide electrode fingers, 20 μm in thickness are separated laterally by a 15 μm gap and placed 25 μm above the Si substrate surface. Since most of the dimensions are scaled by a similar factor between the two geometries, the results of the FE analysis are also qualitatively similar. The simulation was carried out with the central electrode translated downwards in $\Delta(z)$ in steps of 2 μm. The simulation results at $\Delta(z) = 0$ μm and -20 μm are shown in figure 4, and the calculated force density profile as a function of z is shown in figure 5.

In depressing the comb finger towards the substrate, the force density first rises at a rate of 0.0024 $\epsilon_0 V_M^2/\mu m^2$, and the system becomes non-linear below Δz $\sim$ -4 μm. At $\Delta(z)$=0, the calculated force density is 0.00651 $\epsilon_0 V_M^2/\mu m$, only 6% of the value found in the 4×2 μm^2 poly-Si comb fingers. The difference arises from the relative 20% increase in the electrode-substrate gap which decreases the electric field asymmetry between the top and bottom surfaces. The intra-electrode distance is also very much increased to allow for the design rules of electroplated Ni process. The calculated force density per unit length was multlipied by the total length and number of comb fingers, equal to 63.1 mm. Multiplying this by a factor of $\epsilon_0 \times V_M^2$ gives a total net vertical force of 36.4 μN at 100V. This is just 16% larger than the observed value; the difference is likely accounted for by the effects of fringing fields at the ends of each fingers.

The variation of generated force with displacement was investigated by using a capacitive force sensor (FemtoTools FT-S540) which was driven into the central plate of the MEMS device in steps of 100 nm using a motorised stage (PI M-112.1G). The gain of the sensor, and the relative stiffnesses of the sensor and MEMS were calibrated using a Nanoindentation Tester (CSM Instruments), and the measured results adjusted to reflect the fully blocked force of the MEMS actuator [8]. The measurements were taken with a voltage applied to the MEMS of V_M = 50 V, at a frequency of 5 kHz. Both the static and dynamic force was measured at each step in z in order to calculate the actual distance moved by the stage. The calibration was limited by the deflection range of the force sensor, however it is clear to see from the results in figure 6 that the actual increase in applied force is much less than predicted. This is due to the fact that when the central plate of the device is pushed into the substrate, it is the innermost electrodes, which happen to be the shortest within the design, that are displaced by the greatest amount, whereas

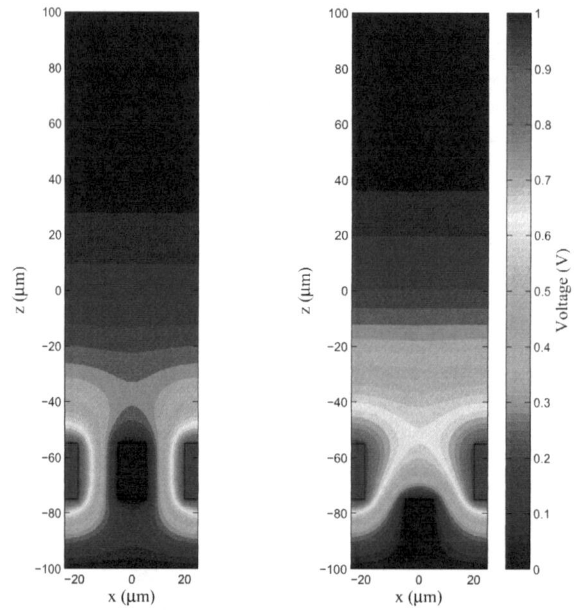

Figure 4: (Colour Online) Calculated potential for the geometry of the device, with the rotor coplanar with the stators (left) and depressed 20 μm below (right). The outline of the electrodes are indicated by the black lines.

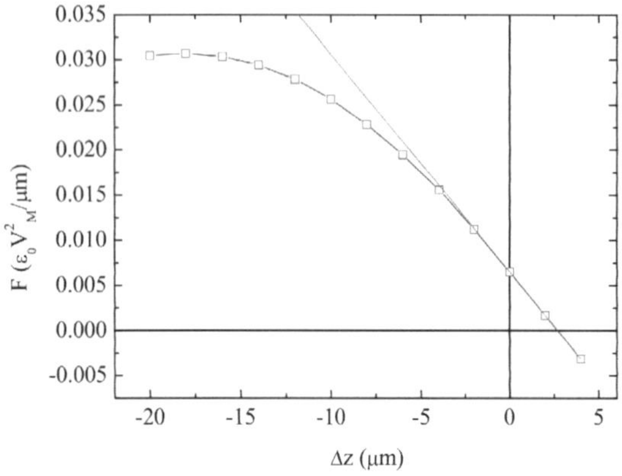

Figure 5: (Colour online) Plot of the variation of net vertical force density calculated by FEA as a function of the vertical displacement of the comb finger (blue squares). The linear behaviour in the region of small displacements is indicated by the fitted pink line.

the longer comb fingers that are closer to the edges of the device are more closely anchored to the substrate.

4. Conclusion

The shaded area in figure 6 represents the accuracy to which it is possible to align the sample and MEMS within an actual piezoelectric measurement. The overall error on the force measurement was slightly increased to represent the small measured variation in force with displacement, in comparison to the accuracy to which the force can be measured with the FemtoTools sensor. However it is clear that the variation in generated force is much less than that predicted by the FEA, for the reason stated above. In fact it is not possible to push the central plate of the device much further towards the substrate without breaking the suspension beams, due to the stiffness and intrinsic stresses within the Ni thick films from which they are made. Further work is

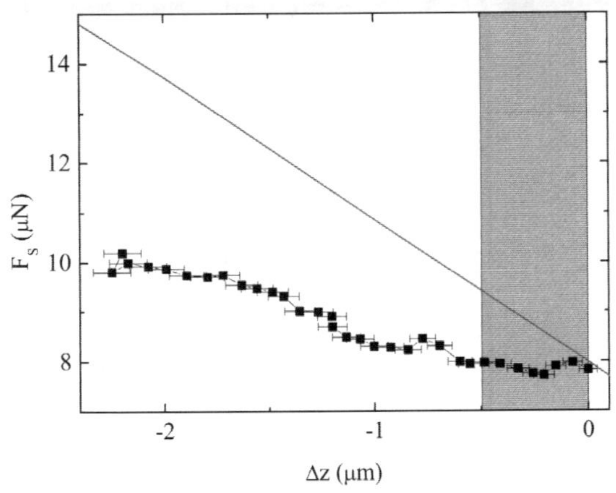

Figure 6: (Colour online) Variation of the measured dynamic force as a function of distance that the central pad of the MEMS device is physically depressed (black squares). The results from the FE modelling presented in figure 5, normalised to the measured force at V_M = 50 V are shown by the pink line. The shaded grey area indicates the range of displacement of the MEMS when in use.

required to re-design the flexural supporting beams to address this issue. Nonetheless from this analysis, provided the optical alignment is carefully carried out during the measurement, we can conclude that the force applied to the micro-scale piezoelectric sample is accurately quantified.

In summary, a MEMS device was produced for the evaluation of piezoelectric coefficients in micro-scale components. The force generated by the electrostatic comb drive actuator was characterised as a function of downwards displacement of the central suspended plate. Finite element analyses of the electric potential around a 2D cross section of the actuator electrodes were successfully used to predict the variation in levitation force. The models showed good correlation with the measured force with zero displacement. The increase in force with decreasing z was found to be smaller than the variation predicted by the models, due to the fact that not all the electrodes are displaced by equal amounts due to the flexural supporting beams of the device which attach the comb fingers to the MEMS substrate.

Acknowledgments

This research was supported by the UK's National Measurement System.

[1] Wilson S A, Jourdain R P J, Zhang Q, Dorey R A, Bowen C R, Willander M, Wahab Q U, Willander M, Safaa M A h, Nur O, Quandt E, Johansson C, Pagounis E, Kohl M, Matovic J, Samel B, van der Wijngaart W, Jager E W H, Carlsson D, Djinovic Z, Wegener M, Moldovan C, Iosub R, Abad E, Wendlandt M, Rusu C and Persson K 2007 *Mater. Sci. Eng. R-Rep.* **56** 1–129
[2] Muralt P 2000 *J. Micromech. Microeng.* **10** 136–146
[3] Zuo C, Van der Spiegel J and Piazza G 2010 *IEEE Trans. Ultrason. Ferroelectr. Freq. Control* **57** 82–87
[4] Piekarski B, DeVoe D, Dubey M, Kaul R and Conrad J 2001 *Sens. Actuator A-Phys.* **91** 313–320
[5] Rebeiz G and Muldavin J 2001 *IEEE Microwave Magazine* **2**(4) 59–71
[6] Wang L, Deng K, Zou L, Wolf R, Davis R and Trolier-McKinstry S 2002 *IEEE Electron Device Lett.* **23** 182–184
[7] Beeby S P, Tudor M J and White N M 2006 *Meas. Sci. Technol.* **17** R175–R195
[8] Wooldridge J, Muniz-Piniella A, Stewart M, Shean T A V, Weaver P M, and Cain M G 2013 *J. Micromech. Microeng.* **23**(3) 035028
[9] Wiak S and Smolka K 2009 *Compel-Int. J. Comp. Math. Electr. Electron. Eng.* **28** 593–602
[10] Guo D and Zhu Y 2009 *Advanced Tribology* ed Luo J, Meng Y, Shao T and Zhao Q pp 566–570
[11] Timpe S J, Hook D A, Dugger M T and Komvopoulos K 2008 *Sens. Actuator A-Phys.* **143** 383–389
[12] Molfese A, Nannini A, Pennelli G and Pieri F 2006 *Analog Integr. Circuits Process.* **48** 33–40
[13] Chyuan S W, Liao Y S and Chen J T 2005 *J. Microelectromech. Syst.* **14** 305–312
[14] Liao Y, Chyuan S and Chen J 2004 *J. Micromech. Microeng.* **14** 1258–1269
[15] Painter C and Shkel A 2004 *Proceedings of the IEEE Sensors 2004, Vols 1-3* IEEE Sensors ed Rocha D, Sarro P and Vellekoop M pp 508–511
[16] van Spengen W M and Heeres E C 2007 *J. Micromech. Microeng.* **17** 447–451
[17] Blackburn J F, Vopsaroiu M and Cain M G 2008 *J. App. Phys.* **104** 074104
[18] Tang W, Lim M and Howe R 1992 *J. Microelectromech. Syst.* **1**(4) 170–8

Absorption current behaviour of polyethylene/silica nanocomposites

K Y Lau[1, 3, 4], A S Vaughan[1], G Chen[1], I L Hosier[1] and A F Holt[2]

[1] School of Electronics and Computer Science, University of Southampton, Southampton SO17 1BJ, United Kingdom
[2] School of Chemistry, University of Southampton, Southampton SO17 1BJ, United Kingdom
[3] Faculty of Electrical Engineering, Universiti Teknologi Malaysia, 81310 Johor Bahru, Malaysia

E-mail: kyl1g10@ecs.soton.ac.uk

Abstract. Absorption current is an important characteristic of polymers with regard to their time-domain response to a direct current (DC) poling field. This is because the results of absorption current measurements can be used to gain understanding of the relationship between space charge accumulation and movement. In semicrystalline polyethylene, for example, charge accumulation is likely to be influenced by the presence of charge trapping sites, which are associated with interfaces between the crystalline and amorphous phases. With the addition of a nanofiller, the charge transport mechanism will become more complicated than in the unfilled polymer, as the inclusion of the nanofiller will introduce nanofiller/polymer interfaces. The presence of such interfaces will affect the current flow due to the introduction or modification of the distribution of trapping sites within the system. In this paper, we report on an investigation into the absorption current behaviour of polyethylene nanocomposites containing 0 wt%, 2 wt%, 5 wt% and 10 wt% of silica nanofiller, either untreated or treated using trimethoxy(propyl)silane coupling agent. Our results indicate that the absorption current behaviour of the polyethylene was affected by the presence of the nanosilica. While the current behaviour through the unfilled polymer decreases with time in a conventional manner, all nanocomposites reveal an initial decrease followed by a period in which the current increases with increasing time of DC field application.

1. Introduction

When an electric field is applied to a non-ideal dielectric material between two plane parallel electrodes, the applied field interacts with charges, causing their motion, which manifests itself as a current flow in the external circuit [1]. The current flow can mainly be categorized into three types, as follows: the initial current that flows through the material is the capacitive charging current, which causes a dramatic rise at the very beginning of the voltage application. This is followed by a gradual decrease of current, known as the absorption current or the anomalous current. Conventionally, the absorption current decreases slowly until it reaches a quasi-steady state, providing a conduction current which is often used to compare the conductivity of different dielectric materials.

[4] To whom any correspondence should be addressed.

The slowly decaying current, i.e. the absorption current, is considered to be an important characteristic of polymers with regard to their time-domain response to a DC poling field. This is because results from absorption current measurements can be related to space charge measurements to gain a better understanding of relationship between space charge accumulation and movement, as well as the resulting effects in the external circuit [2]. In general, factors affecting the absorption current include electrode polarization, dipole orientation, charge accumulation, tunnelling of charge carriers from the electrodes to empty traps and hopping of charge carriers through localized states [3].

The determination of charge transport mechanisms in polymeric insulation is, however, complicated in comparison with conducting and semiconducting materials. In semicrystalline polyethylene, for example, the crystallites are surrounded by amorphous regions, and there are likely to be high concentrations of traps at the crystalline/amorphous interfaces. With the addition of a filler, the charge transport mechanism is likely to become more complicated than in the unfilled polymer. In nanocomposites, for example, the inclusion of nanoparticles will introduce additional interfaces between the nanofiller and the polymer, as has been emphasized by many researchers [4-7]. The presence of such nanofiller/polymer interfaces will affect the current flow due to the introduction of additional nanofiller/polymer trapping sites and/or through the modification of the original crystalline/amorphous trapping sites. Therefore, investigations into absorption current behaviour would contribute to a better understanding of filler/polymer interactions in nanocomposites, in particular, in relation to charge transport mechanisms.

2. Experimental

2.1. Materials and Preparation

The polymers used in this study were the low density polyethylene (LDPE) grade LD100BW, obtained from ExxonMobil Chemicals, and the high density polyethylene (HDPE) grade Rigidex HD5813EA, obtained from BP Chemicals. The nanofiller used was silicon dioxide (SiO_2) nanopowder (nanosilica) obtained from Sigma-Aldrich, with a quoted particle size range of 10 nm to 20 nm.

Surface treatment of the nanosilica was performed using trimethoxy(propyl)silane via an anhydrous route, while the nanocomposites were prepared using a solution blending method; these systems are subsequently described as C3-treated. Details of the nanosilica surface treatment and the nanocomposite sample preparation procedure can be found elsewhere [8]. The required test samples (~200 μm thick) were melt pressed at 150 °C, followed by isothermal crystallization in an oil bath at 115 °C for 1 h. The samples were sputtered coated with gold to ensure good electrical contact between the electrodes and the samples.

2.2. Characterization

Absorption current measurements were performed using a Keithley 6487 picoammeter / voltage source and a sample holder with two opposing 20 mm diameter electrodes. A step voltage equating to an applied field of 40 $kV_{(DC)}$ mm^{-1} was applied and the absorption current was recorded. Prior to measurement, each sample was grounded at 60 °C in a vacuum for at least 72 h so that excess charge would dissipate.

3. Results

3.1. Absorption Current Measurements at 40 kV mm^{-1} DC Field

Figure 1 shows the plot of the time dependence absorption current for all investigated samples at a constant DC field of 40 kV mm^{-1} over 1 h at room temperature. It is obvious that all the nanocomposites exhibit a current-time characteristic that is different from that of the unfilled polyethylene. At the beginning of the test, it can be noticed that the rate of decrease of current for all the nanocomposites is significantly greater than for the unfilled polyethylene (see the fitted lines).

It is noteworthy that, although the absorption current values differ for each sample at the beginning of the test, these variations fall within experimental uncertainties; repeated experimental runs showed that data variations within a factor of 3 are typical, and can occasionally be up to a factor of 5, which causes the exact quantitative analysis of such behaviour to become questionable. The poor reproducibility of test data is, however, a common issue with absorption current measurements [1]. Quantitative analysis of this facet of the data is therefore not the main interest of this study.

After the initial relatively rapid decrease of current, all the nanocomposites exhibit a characteristic reduction in slope (at ~200 s). The precise time at which this occurs varies from sample to sample, even if the same type of sample were used. No comparable change of slope occurs in the unfilled polyethylene.

Finally, at long times, the current flowing through the nanocomposites begins to rise (at ~1000 s). Such behaviour is, again, not found in the unfilled polyethylene. To aid with data interpretation, Figure 2 was plotted to show the current-time characteristics of all the investigated samples, obtained over a longer time scale (~3 h). From this, the current behaviour of the nanocomposites can be interpreted based on three phases: the region corresponding to the initial decrease of current (Phase I), the region immediately following the initial reduction in slope (Phase II) and the region where the current rises (Phase III).

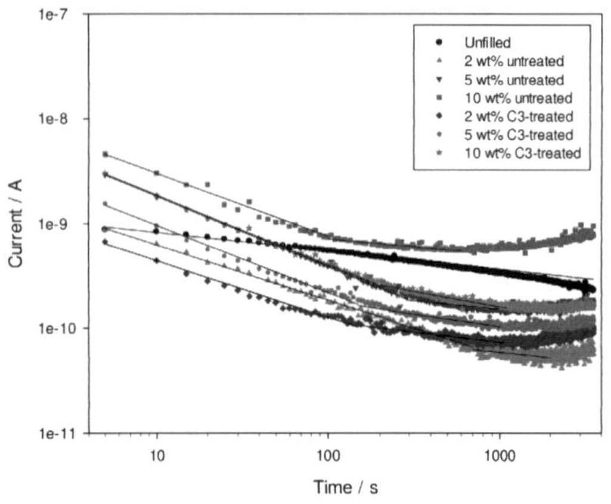

Figure 1. Plot of absorption current against time up to 1 h for all investigated samples at an applied field of 40 kV mm^{-1}. Two lines are fitted for nanocomposites to indicate a change in slope prior to a current rise.

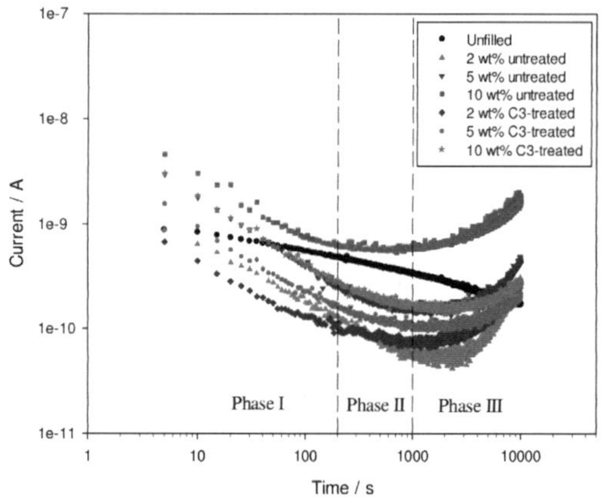

Figure 2. Plot of absorption current against time up to 10^4 s for all investigated samples at an applied field of 40 kV mm^{-1}. The data were divided into three phases for the ease of interpretation.

3.2. Data Analysis

The current-time characteristic of the absorption current is often found to follow the power law relationship:

$$I = At^{-b} \tag{1}$$

where I is the current, t is the time after the application or removal of the external voltage, A is a temperature dependent factor and b is a constant representing the slope of the log-log current-time plot. It should be noted that simple interpretation of the exponent b is complicated by the fact that polymers exhibit a distribution of relaxation times and that these times may not be easily determined in polyethylene at normal temperatures [2].

In Phase I, although the absolute magnitude of the absorption current was found to vary from sample to sample, repeated experimental runs showed the rate of decrease of current with time to be reproducible in all the investigated nanocomposites. Significantly, the rate at which the current decreased was found to be significantly greater at the beginning of the test in comparison with the unfilled polyethylene; the variation of current with poling time at the beginning of the test (0 s to 200 s) is summarized in Table 1.

Table 1. Exponent calculated from the absorption current data at the beginning of the test (0 s to 200 s) at an applied DC field of 40 kV mm^{-1}. The symbol b_I was used specifically to refer to exponent b in Phase I.

Sample	$b_I \times 10^{-2}$
Unfilled	17.5 ± 0.3
2 wt% untreated	52.7 ± 0.6
5 wt% untreated	67.3 ± 0.4
10 wt% untreated	59.6 ± 1.3
2 wt% C3-treated	53.9 ± 1.0
5 wt% C3-treated	64.4 ± 0.9
10 wt% C3-treated	66.7 ± 0.6

Generally, all the nanocomposites possess a higher value of the exponent b_I in Phase I, when compared with the unfilled polyethylene. Also, this parameter increases with increasing amount of the nanosilica, although any influence of nanosilica surface treatment appears to be comparable to experimental uncertainties. Figure 3 shows a plot of exponent b_I against nanosilica content; the indicated b_I values of the nanocomposites were obtained from the average values of the untreated and C3-treated systems.

Using the data in Figure 1, the charge carrier mobility can be estimated in each system using the formula [9]:

$$\mu = \frac{0.786d^2}{t_p V} \tag{2}$$

where μ is the mobility of charge carriers, d is the sample thickness, t_p is the time at which the absorption current is at its maximum (or the time at which a slope change occurs) and V is the applied voltage. The resulting charge carrier mobility is shown in Figure 4; the change of slope for the unfilled polyethylene was assumed to occur at 10000 s. It can be noticed that all the nanocomposites are characterized by a higher charge carrier mobility than the unfilled polyethylene; the charge carrier mobility estimated for the unfilled polyethylene falls within the measurement range of a series of polyethylenes reported by Montanari et al. [10], suggesting that our analysis is reasonable.

In Phase II, the exact quantitative determination of the exponent b was complicated somewhat by the difficulty in establishing the appropriate time range. Despite this, the same fitting procedure described above was, nevertheless, employed and the resulting estimated b_2 values are shown in Table 2. Comparing the data presented in Table 1 and Table 2, it is evident that the rate at which the current decreases for all the nanocomposites in Phase II is much lower than in Phase I; this does not apply to the unfilled polyethylene. While the quantitative value of exponent b_2 in Phase II is not crucial for our argument, we believe it is important to note that the values become comparable to that which characterizes the unfilled polyethylene. It is noteworthy that a comparable quantitative analysis was not attempted for Phase III since the observation of current rise is not common in the context of absorption current measurements.

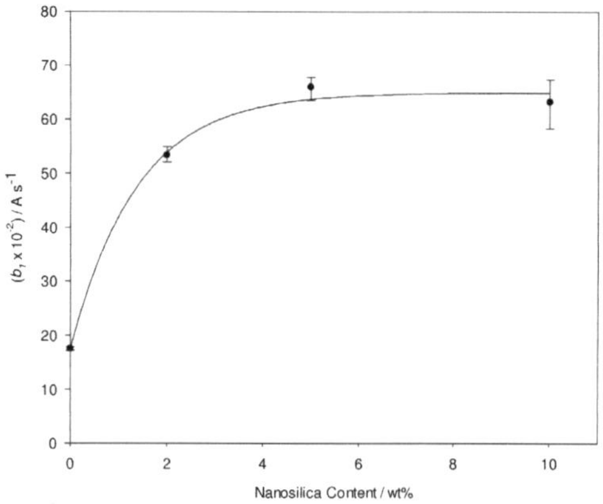

Figure 3. Plot showing the variation of the exponent b_1 as a function of nanosilica content. The b_1 values of the nanocomposites were obtained from the average values of the untreated and C3-treated systems.

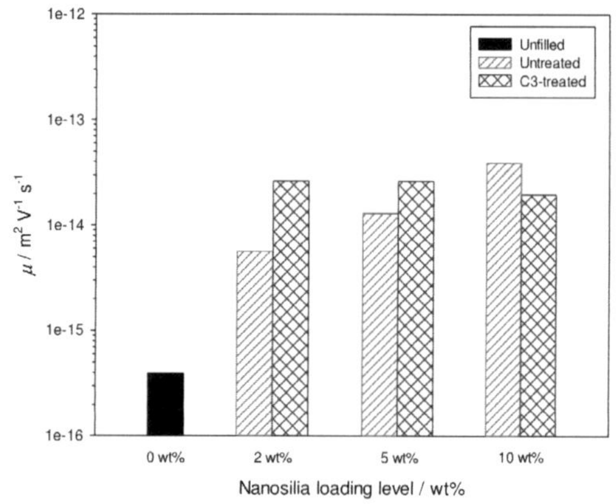

Figure 4. Charge carrier mobility of unfilled polyethylene and nanocomposites containing different types and amounts of nanosilica.

Table 2. Exponent calculated from the absorption current data in Phase II. The symbol b_2 was used specifically to refer to exponent b in Phase II.

Sample	$b_2 \times 10^{-2}$
Unfilled	17.5 ± 0.3
2 wt% untreated	17.3 ± 2.0
5 wt% untreated	16.5 ± 2.4
10 wt% untreated	14.6 ± 1.3
2 wt% C3-treated	19.6 ± 0.9
5 wt% C3-treated	24.7 ± 0.8
10 wt% C3-treated	30.2 ± 0.8

4. Discussion

In Phase I, the current decreases more rapidly in the nanocomposites than in the unfilled polyethylene. This implies that the incorporation of nanosilica into polyethylene has increased the mobility of charge carriers in the nanocomposites, so leading to faster decay of the current as the carriers take a shorter time to reach the electrodes. It is noteworthy that $0 \leq b \leq 2$ is consistent with dipole orientation, carrier tunnelling and carrier hopping while $0 \leq b \leq 1$ is consistent with charge injection forming trapped space charge [11]. While nanoparticles with their large surface areas may act as additional electron traps, they may also reduce the average hopping distance relative to that of the polymer matrix and thus increase the mobility [12]. As the nanofiller content increases, charge mobility may increase due to increased nanofiller/polymer interfaces. This is consistent with the analysis of b_1 in Phase I, where this parameter increases with the nanofiller content.

The mobility of charge carriers is highly dependent upon the release of charges from deep traps in addition to the underlying current caused by the large number of shallower traps introduced by the nanoparticles [2]. Analysis of charge carrier mobility shows that all the nanocomposites possess higher charge mobility in comparison with the unfilled polyethylene. Although it remains difficult to judge the effect of nanofiller type and amount on the charge carrier mobility, it is reasonable to deduce that, the incorporation of nanosilica into polyethylene results in the presence of shallower traps that are related to the nanofiller/polymer interfaces in addition to the original trap distribution that characterizes the crystalline/amorphous interfaces; the former serves to assist in charge transport.

The change in the *b* exponent seen in all the nanocomposites, on entering Phase II, could indicate a change in dominant mechanism of the absorption current. This could be attributed to interfacial polarization, as explained elsewhere in the literature [13-14]. The change of slope could also be interpreted in the following way: electronic transport is effectively controlled by trapping, i.e., an electron may travel rapidly through the system for a short time, but its effective or average mobility is greatly reduced as a result of being immobilized for much longer period in localized states (traps) [11]. Alternatively, the change in slope could indicate that the absorption current in Phase II is dominated by the matrix rather than the nanofiller (which we believe dominates in Phase I), since the decay in absorption current within this phase occurs at a rate that is: i) very different from Phase I and ii) comparable to that seen in the unfilled polyethylene.

Although the increasing absorption current with poling time seen in Phase III is extremely unusual, such observations are not without precedent. A comparable increase of current was also discovered by Smith [2] for microcomposites and was thought to be caused by interfacial polarization related to the interfaces between the filler and the matrix. However, no convincing explanation has been proposed in relation to this phenomenon.

5. Conclusions

While the current behaviour through the unfilled polyethylene decreases with time in a conventional manner, all nanocomposites reveal an initial decrease followed by a period in which the current increases with increasing time of DC field application. In addition, the inclusion of nanosilica into polyethylene causes the decay of the absorption current initially to occur more rapidly than in the unfilled PE, highlighting the possibility of increased charge mobility in the nanocomposites. Presently, the literature concerning such current-time characteristics of absorption current is scarce and more work is necessary fully to explain the underlying charge transport mechanisms in nanocomposites.

6. References

[1] Adamec V and Calderwood J H 1978 *J. Phys. D: Appl. Phys.* **11** 781-800
[2] Smith R C 2009 *Mechanistic Electrical Behaviour of Crosslinked Polyethylene / Silica Nanocomposites* Ph.D. Thesis Rensselaer Polytechnic Institute
[3] Das Gupta D K and Joyner K 1976 *J. Phys. D: Appl. Phys.* **9** 829-40
[4] Nelson J K 2007 *Proc. Electrical Insulation Conf. and Electrical Manufacturing Expo. (Nashville)* pp 229-35
[5] Lewis T J 2004 *IEEE Trans. Dielectr. Electr. Insul.* **11** 739-53
[6] Tanaka T, Kozako M, Fuse N and Ohki Y 2005 *IEEE Trans. Dielectr. Electr. Insul.* 2005 **12** 669-81
[7] Raetzke S and Kindersberger J 2010 *IEEE Trans. Dielectr. Electr. Insul.* **17** 607-14
[8] Lau K Y, Vaughan A S, Chen G, Hosier I L and Holt A F 2013 *J. Phys. D.: Appl. Phys.* **46** 095303
[9] Many A and Rakavi G 1962 *Physical Review* **126** 1980-8
[10] Montanari G C, Mazzati G, Palmieri F, Motori A, Perego G and Serra S 2001 *J. Phys. D.: Appl. Phys.* **34** 2902-11
[11] Wintle H J 1983 Conduction processes in polymers *Engineering Dielectrics* vol IIA ed R Bartnikas and R M Eichhorn (Philadelphia: ASTM) pp 239-354
[12] Fleming R J 2010 The emerging mechanistic picture *Dielectric Polymer Nanocomposites* ed J K Nelson (New York: Springer) pp 285-319
[13] Roy M, Nelson J K, MacCrone R K and Schadler L S 2007 *J. Mater. Sci.* **42** 3789-99
[14] Roy M 2005 *An Examination of the Potential for Nano-composites in the Formulation of HV Cable Insulation* Ph.D. Thesis Rensselaer Polytechnic Institute

Acknowledgements

The authors acknowledge the Ministry of Higher Education, Malaysia and Universiti Teknologi Malaysia for financial sponsorship.

Development of PLZT film-on-foil capacitors with high energy density

Beihai Ma, Manoj Narayanan, Shanshan Liu, Zhongqiang Hu and Uthamalingam Balachandran

Energy Systems Division, Argonne National Laboratory, Argonne, IL 60439, USA

Email: bma@anl.gov

Abstract. Ceramic thin films with high dielectric constant and high breakdown strength hold special promise for advanced electronics applications. We deposited lanthanum-doped lead zirconate titanate (PLZT) on base metal foils (film-on-foils) by chemical solution deposition. The PLZT film-on-foils were characterized over a wide temperature range between room temperature and 200°C. For ≈2-μm-thick PLZT films grown on $LaNO_3$-buffered nickel foils at room temperature, we measured a dielectric constant of ≈1300, dielectric loss of ≈0.07, remanent polarization of ≈26 μC/cm^2, coercive field of ≈29 kV/cm, leakage current density of ≈9.2 × 10^{-9} A/cm^2, and energy density of ≈67 J/cm^3. Dielectric breakdown strength was determined to be ≈2.6 × 10^6 V/cm by Weibull analysis. These and other results reported here indicate that electronic devices with film-on-foil capacitors would possess higher performance, improved reliability, and enhanced volumetric and gravimetric efficiencies compared with traditional dielectric capacitors.

1. Introduction

With development of green energy technologies and their expanding applications, devices for effectively absorbing, converting, storing, and supplying electrical energy are in increasing demand. Not only high energy storage density but also high-power output is required for many applications. Commercial electrical energy storage and supply devices include fuel cells, batteries, electrochemical supercapacitors, and dielectric capacitors. Among them, fuel cells and batteries exhibit high energy density, but relatively low power density because of the slow movement of charge carriers. Electrochemical supercapacitors offer improved power density at moderate energy density and are promising for many power system applications, but their charge and/or discharge process still requires seconds or tens of seconds. In contrast, dielectric capacitors provide the highest power density because of their extremely high charge/discharge speed. This characteristic makes dielectric capacitors well suited to energy storage applications when high power delivery or uptake is required. However, traditional dielectric capacitors possess low energy density (≤0.1 J/cm^3).

Recently, dielectric film capacitors have attracted increased attention because of their applications as decoupling and DC-link capacitors. They can also be used in AC/DC convertors and in portable power electronics, where the capacitors must be able to operate under high voltage and yet have minimal weight and volume [1-2]. Although high-k oxide film dielectrics exhibit high permittivity and breakdown strength (3 to 12 MV/cm) [3], their dielectric constant is relatively low, and in turn, the volumetric energy density is only in the neighborhood of 1 to 2 J/cm^3 [4]. Recently, polymer ferroelectrics, ceramic ferroelectrics, antiferroelectrics, and relaxor ferroelectrics have shown better potential and higher energy density than linear dielectrics [2,5-8]. Chu et al. [2] reported high dielectric breakdown strength and high energy density of ≈17 J/cm^3 in the copolymer system poly(vinylidene fluoride-trifluoroethylene) P(VDF–TrFE), where the introduction of defects in this system converts it to a relaxor ferroelectric with almost negligible remanent polarization. Yao et al. [5] reported relaxor-antiferroelectric thin films, relaxor-ferroelectric thin films, and poly(vinylidene fluoride)-based polymer blend thin films that show high energy density (10 to 25 J/cm^3). Ma et al. [6,7] reported energy density of ≈53 J/cm^3 in antiferroelectric PLZT films grown on metal foils. Hao et al. [8] reported an energy density of ≈30 J/cm^3 in relaxor ferroelectric PLZT films grown on platinized silicon substrates.

By far, the vast majority of ceramic dielectric films are fabricated on single crystals or silicon wafers [9-11]. Use of metallic substrates, however, is more desirable for the fabrication of large-area devices with low cost. Metal foils coated with ceramic dielectric material, the so-called "film-on-foil dielectric sheets," can be laminated onto printed circuit boards (PCBs) to create embedded capacitors that have broad applications in power electronics. Alternatively, these film-on-foils could be stacked to form a capacitor cartridge with minimal footprint on a circuit board. However, substantial technical challenges still exist in the fabrication of film-on-foil capacitors that exhibit high dielectric strength and are crack free. Problems include thermal expansion mismatch between films and metal substrates, formation of a low-dielectric-constant parasitic layer at the film/substrate interface, and diffusion of cations from the substrate into the dielectric film.

We earlier reported fabrication of ferroelectric PLZT film-on-foil capacitors on base metal (nickel and copper) sheets [12-14]. To produce film-on-foil sheets with high quality, we have found it necessary to apply a conductive oxide buffer to the nickel foil prior to the deposition of PLZT. This architecture allows crystallization of dielectric ceramic films grown on nickel foils in air without the formation of deleterious interfacial secondary phases [12]. An alternative approach is to process the ceramic dielectric films in a controlled atmosphere with reduced oxygen partial pressure [14]. In this paper, we report our recent results from dielectric characterization under high bias field for PLZT film-on-foil capacitors grown on LNO-buffered nickel substrates.

2. Experiment

High purity (99.98% pure) nickel substrates with dimensions of 25 mm × 25 mm × 0.4 mm were obtained from MTI Corp. (Richmond, California). They were polished by chemical-mechanical planarization (CMP). A root-mean-square surface roughness of ≈2 nm was measured by atomic force microscopy (AFM) in the tapping mode with 5 μm × 5 μm scan size. Prior to being coated, CMP-polished nickel substrates were ultrasonically cleaned in distilled water, and then wipe-cleaned with acetone and methanol in sequence.

Prior to the PLZT ($Pb_{0.92}La_{0.08}Zr_{0.52}Ti_{0.48}O_3$) deposition, a conductive oxide film of $LaNiO_3$ (LNO) was coated on the Ni substrate by chemical solution deposition to serve as a buffer layer. LNO precursor solutions with 0.2 M concentration were prepared by dissolving an appropriate amount of lanthanum nitrate hexahydrate and nickel acetate tetrahydrate in 2-methoxyethanol (all from Sigma-Aldrich) and refluxing for 2 h inside a chemical glove box. PLZT precursor solutions with 0.5 M concentration were prepared by a modified 2-methoxyethanol synthesis route [12,15] using an appropriate amount of titanium isopropoxide, zirconium n-propoxide, lead acetate trihydrate, and lanthanum nitrate hexahydrate (all from Sigma-Aldrich). The resulting stock solution contained 20% excess lead to compensate for lead loss during the heat treatments described below. Before being used for coating, the LNO and PLZT precursor solutions were filtered through Restek polytetrfluoroethylene (PTFE) syringe filters (Restek Corp., Bellefonte, PA) with 0.22-μm open pore size. The filtered LNO precursor solution was spin coated with a Laurell WS400 spin processor (Laurell Technologies, North Wales, PA) at 3000 rpm for 30 s on the Ni substrates, pyrolyzed at 450°C for 5 min, and annealed at 625°C for 2-5 min in air. This process was repeated three times to build the desired ≈0.4-μm-thick LNO buffer film. Subsequently, filtered PLZT precursor solution was spin coated on the LNO-buffered Ni substrates at 3000 rpm for 30 s, followed by pyrolysis at 450°C for 5 min and annealing at 650°C for 5-10 min for each coating. After every three layers of coating, additional annealing was performed at 650°C for 15 min. Solution coating and firing were repeated to produce films of desired thickness. All pyrolysis and annealing were performed in air in Lindburg tube furnaces. Each coating resulted in a PLZT film of ≈0.115-μm thickness after pyrolysis and crystallization. Film thickness was determined from cross-sectional images with scanning electron microscopy (SEM) [12]. Phase development in the processed films was studied by a Bruker AXS D8 diffraction system. Film microstructures were characterized by a Hitachi S-4700-II field-emission electron microscope. The grain size and surface roughness were determined by a Veeco Instruments D3100 scanning probe microscope operated in the tapping mode.

Platinum (Pt) top electrodes with thickness of 100 nm were deposited by electron-beam evaporation through a shadow mask to define ≈250-μm-diameter capacitors. Samples with Pt top electrodes were annealed at 450°C in air for 2 min for electrode conditioning. A Signatone QuieTemp® probe system with heatable vacuum chuck (Lucas Signatone Corp., Gilroy, CA) was used for electrical characterization. For the electrical measurements, the Pt/PLZT/LNO/Ni heterostructure was contacted by a Pt top electrode pad with one probe and the substrate (bottom electrode) with the other. A positive applied voltage corresponds to a configuration where the top electrode is at a higher potential than the bottom electrode. An Agilent E4980A precision LCR meter was used to determine the capacitance and dissipation factor under an applied bias field. A Radiant Technologies Precision Premier II tester measured the hysteresis loops. The capacitor samples were immersed in Fluka silicone oil (Sigma-Aldrich) during measurements of high-field hysteresis loops and dielectric breakdown. A Keithley 237 high-voltage source meter measured the current-voltage characteristics. The leakage current density was determined by fitting the current density relaxation data to the Curie-von Schweidler equation [16].

3. Results and discussion

Figure 1 shows the X-ray diffraction (XRD) pattern measured on a PLZT film deposited on LNO-buffered Ni substrate. The diffraction pattern indicates that PLZT film is well crystallized and exhibits no preferred crystallographic orientation. The PLZT films exhibit a perovskite structure with tetragonal distortion (JCPDS No. 56-0900). Peaks from LNO buffer and nickel substrate are also visible on the diffraction pattern. No crack or delamination was observed from SEM [12]. As shown in the Fig. 1 inset, the PLZT film has an average grain size of ≈45 nm as determined from atomic force microscopy (AFM).

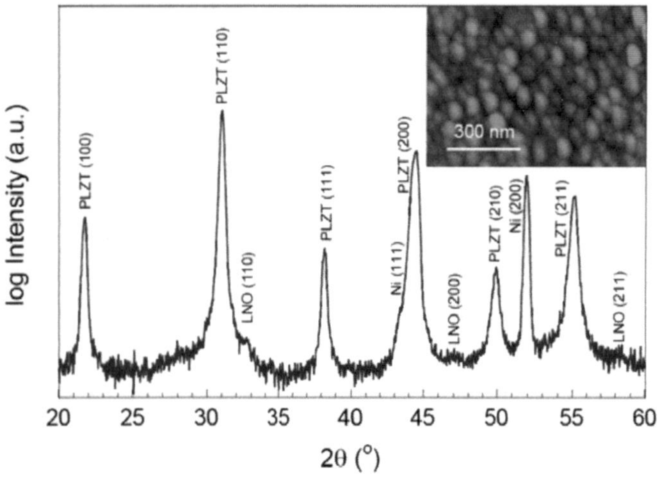

Fig. 1. X-ray diffraction pattern and AFM image (inset) of ≈2-μm-thick PLZT film deposited by chemical solution deposition on LNO-buffered nickel substrate.

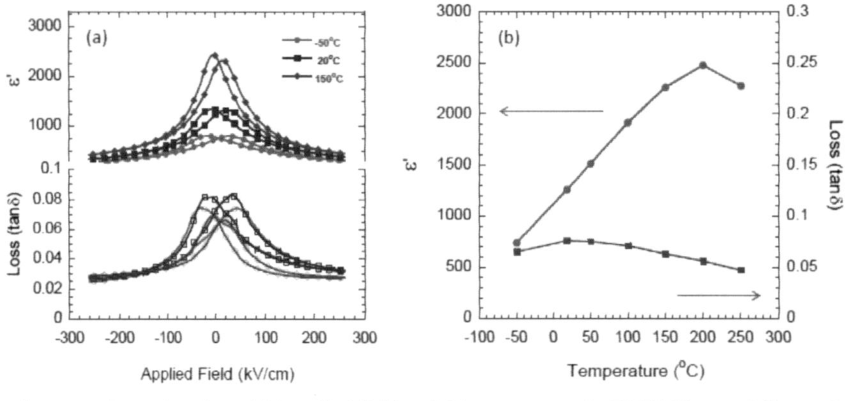

Fig. 2. Dielectric properties as function of (a) applied field and (b) temperature for PLZT film-on-foil capacitor.

Figure 2a shows the dielectric constant (relative permittivity) and loss as a function of applied bias field at three temperatures on a PLZT film grown on LNO-buffered nickel substrate. At zero applied field, we measured dielectric constant of ≈700 and dielectric loss of ≈0.06 at -50°C, dielectric constant of ≈1300 and loss of ≈0.07 at 20°C, and dielectric constant of ≈2200 and dielectric loss of ≈0.05 at 150°C. The dielectric constant increases while dielectric loss decreases with increasing temperature. At 200°C, we measured a dielectric constant of 2500 and loss of ≈0.05. Both dielectric constant and loss decrease with applied bias field. A dielectric tunability, 1 - $\varepsilon'(E)/\varepsilon'(0)$, of ≈55% was measured at field of 100 kV/cm. Figure 2b shows the dielectric properties of a

PLZT/LNO/Ni sample measured within the temperature range between -50 and 250°C. The dielectric constant increases while the dissipation factor decreases with increasing temperature from room temperature to 200°C. Maxima of the dielectric constant and dissipation factor occurred at ≈200°C and room temperature, respectively.

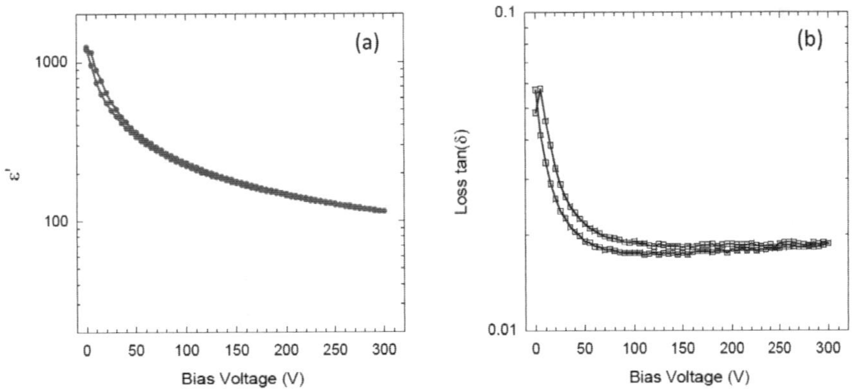

Fig. 3. Bias voltage dependency of (a) dielectric constant and (b) loss tangent of PLZT/LNO/Ni film-on-foil capacitor at room temperature and 10 kHz.

Figure 3 shows dielectric constant and loss measured at room temperature and 10 kHz under bias voltage up to 300 V. We measured a dielectric constant of ≈110 and dielectric loss ≈0.02 under applied bias voltage of 300 V. This PLZT film of ≈2-μm thickness was subjected to a bias field of ≈1.5 × 10^6 V/cm when a 300-V bias voltage was applied. The charging and discharging curves nearly overlay under high bias. The dielectric constant decreases steadily with increasing bias field because of the ferroelectric nature of the PLZT film. Because dipole movement is clamped under high bias field [17], the apparent dielectric constant is depressed with increased field. The dielectric loss drops rapidly under applied bias voltage up to ≈50 V; and it remains nearly constant at ≈0.02 with further increase in bias voltage.

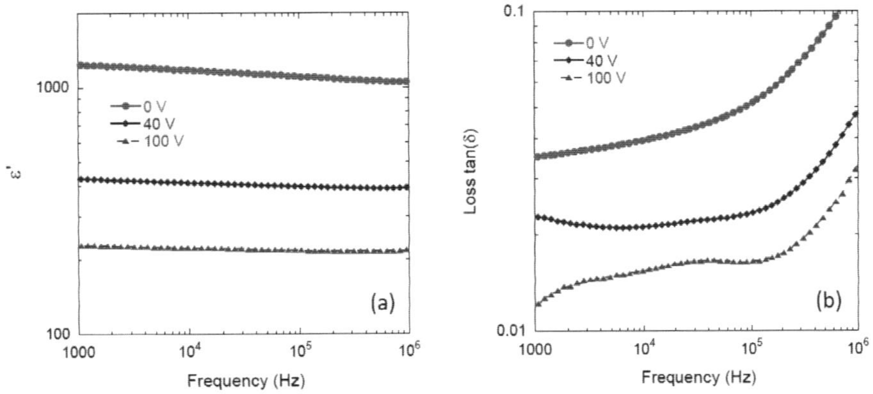

Fig. 4. Frequency dependency of (a) dielectric constant and (b) loss tangent for PLZT/LNO/Ni film-on-foil capacitor under different bias voltages at room temperature.

Figure 4 shows the dielectric constant and dielectric loss measured as a function of frequency under various applied bias voltages up to 100 V. The dielectric constant decreases slightly while the dielectric loss increases substantially with increasing frequency. At 100 kHz under 100 V bias, we measured a dielectric constant of ≈220 and dielectric loss ≈0.02 at room temperature.

Figure 5 shows polarization-electric field (P-E) hysteresis loops of a ≈2-μm-thick PLZT grown on LNO-buffered nickel foil measured with sweeping frequency of 100 Hz under low and high applied field conditions. From the P-E hysteresis loop, we measured remanent polarization P_r = 25.7 μC/cm^2 and coercive field E_c = 28.9 kV/cm, as shown in Fig. 5a. From the high-field hysteresis loop, we measured a recoverable energy density of ≈67 J/cm^3 with a maximum applied voltage of 800 V (corresponding to an applied field of ≈4.0 × 10^6 V/cm), as shown in Fig. 5b. The ratio of recoverable energy to input energy is ≈75%.

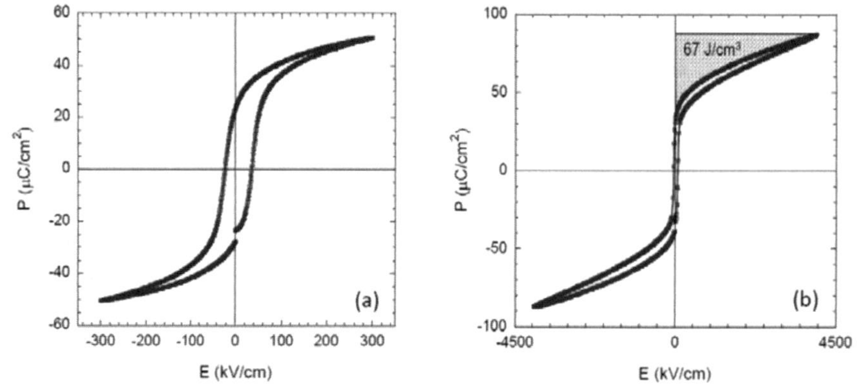

Fig. 5. P-E hysteresis loops for PLZT/LNO/Ni film-on-foil capacitor measured with (a) low field up to 60 V and (b) high field up to 800 V.

Figure 6 shows P_r and E_c measured as a function of temperature with low field condition. Both P_r and E_c decrease steadily with increasing temperature. We measured P_r = 25.7 μC/cm^2 and E_c = 28.9 kV/cm at room temperature, as well as P_r = 7.5 μC/cm^2 and E_c = 8.2 kV/cm at 200°C, for a ≈2-μm-thick PLZT film grown on LNO-buffered nickel foil.

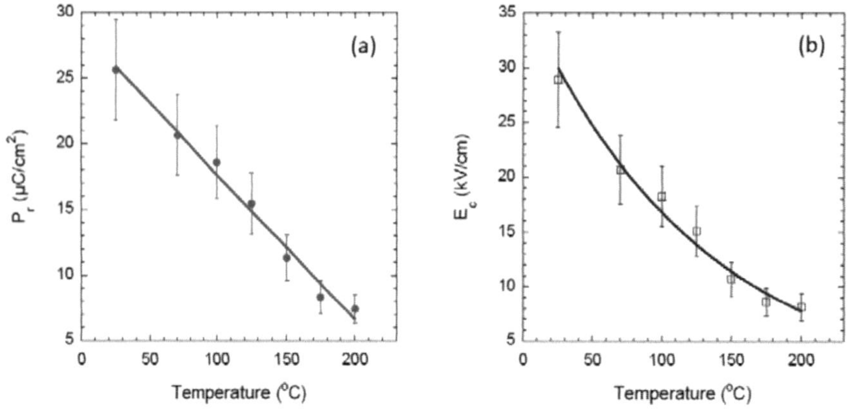

Fig. 6. Temperature-dependent (a) remanent polarization and (b) coercive field of PLZT/LNO/Ni film-on-foil capacitor.

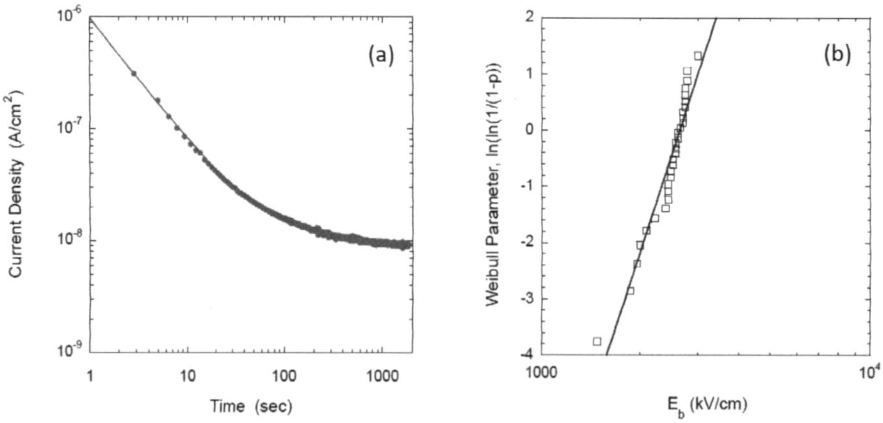

Fig. 7. (a) Time-relaxation current density and (b) Weibull plot measured on PLZT/LNO/Ni film-on-foil capacitor at room temperature.

Figure 7a shows the time relaxation for the current density measured at room temperature on PLZT/LNO/Ni with a constant bias field of ≈100 kV/cm across the top and bottom electrodes. The measurements were conducted by keeping the top Pt electrode positive and the bottom Ni electrode grounded. The relaxation curve shows strong initial time dependence. Fitting data to the Curie-von Schweidler law [16], we determined steady-state leakage current densities of $\approx 9.2 \times 10^{-9}$ A/cm^2 for ≈2-µm-thick PLZT grown on LNO/Ni substrate. Figure 7b shows the Weibull plot of breakdown field strength obtained at room temperature from 30 capacitor samples of PLZT grown on LNO/Ni substrates. The solid straight line is a least-square fitting to the two-parameter distribution function [18,19]. We measured mean breakdown field of 2.6×10^6 V/cm and Weibull modulus β of 7.8 for ≈2-µm-thick PLZT grown on LNO/Ni substrates.

4. Conclusions

We have grown high-breakdown-strength ferroelectric PLZT films on LNO-buffered Ni foils by chemical solution deposition. X-ray diffraction revealed that the PLZT films are well crystallized after heat treatment at 650°C. The dielectric constant increases while the dielectric loss decreases with increasing temperature from room temperature to 200°C. We measured dielectric constant of ≈700 and loss of ≈0.06 at -50°C, dielectric constant of ≈1300 and loss of ≈0.07 at 20°C, and dielectric constant of ≈2200 and loss of ≈0.05 at 150°C, respectively. Hysteresis measurement revealed that both remanent polarization P_r and coercive field E_c decrease with increasing temperature. We measured P_r of ≈26 µC/cm^2 and E_c of ≈29 kV/cm at 20°C, and P_r of ≈7.5 µC/cm^2 and E_c of ≈8.2 kV/cm at 200°C. Steady-state leakage current densities of $\approx 9.2 \times 10^{-9}$ A/cm^2, mean breakdown field of $\approx 2.6 \times 10^6$ V/cm, and energy density of ≈67 J/cm^3 were measured at room temperature.

Acknowledgment

This work was funded by the U.S. Department of Energy, Vehicle Technologies Program, under Contract DE-AC02-06CH11357.

References

[1] S. A. Sherrill, P. Banerjee, G. W. Rubloff, and S. B. Lee, *Phys. Chem. Chem. Phys.* **13** (2011) 20714.
[2] B. Chu, X. Zhou, K. Ren, B. Neese, M. Lin, Q. Wang, F. Bauer, and Q. M. Zhang, *Science* **313** (2006) 334.
[3] J. W. McPherson, J. Kim, A. Shanware, H. Mogul, and J. Rodriguez, *IEEE Trans. Electron Dev.* **50** (2003) 1771.
[4] N. Ortega, A Kumar, J F Scott, D. B. Chrisey, M. Tomazawa, S. Kumari, D. G. B. Diestra, and R S Katiyar, *J. Phys.: Condens. Matter* **24** (2012) 445901.
[5] K. Yao, S. Chen, M. Rahimabady, M. S. Mirshekarloo, S. Yu, F. E. H. Toy, T. Sritharan, and L. Lu, *IEEE Trans. Ultrason. Ferroelectr. Freq. Control* **58** (2011) 1968.
[6] B. Ma, D. K. Kwon, M. Narayanan, U. Balachandran, *Mater. Res. Bull.* **44** (2009) 11.
[7] B. Ma, D. K. Kwon, M. Narayanan, U. Balachandran, *J. Mater. Res.* **24** (2009) 2993.
[8] X. Hao, Y. Wang, J. Yang, S. An, and J. Xu, *J. Appl. Phys.* **112** (2012) 114111.
[9] H.-J. Zhao, T.-L. Ren, N.-X. Zhang, R.-Z. Zuo, X.-H. Wang, L.-T. Liu, Z.-J. Li, Z.-L. Gui, and L.-T. Li, *Mater. Sci. Eng. B* **99** (2003) 195.
[10] L. B. Kong and J. Ma, *Mater. Lett.* **56** (2002) 30.
[11] S. Chao, B. Ma, S. Liu, M. Narayanan, and U. Balachandran, *Mater. Res. Bull.* **47** (2012) 907.
[12] B. Ma, D. K. Kwon, M. Narayanan, and U. Balachandran, *J. Electroceram.* **22** (2009) 383.
[13] U. Balachandran, D. K. Kwon, M. Narayanan, and B. Ma, *J. Europ. Ceram. Soc.* **30** (2010) 365.
[14] M. Narayanan, B. Ma, and U. Balachandran, *Mater. Lett.* **64** (2010) 22.
[15] Q. Zou, H. E. Ruda, and B. G. Yacobi, *Appl. Phys. Lett.* **78** (2001) 1282.
[16] K. Jonscher, *Dielectric Relaxation in Solids*, Chelsea Dielectrics Press, London (1983).
[17] B. Ma, Z. Hu, S. Liu, M. Narayanan, and U. Balachandran, *Appl. Phys. Lett.* **102** (2013) 072901.
[18] W. Weibull, *J. Appl. Mech.* **18** (1951) 293-297.
[19] L. A. Dissado, *J. Phys. D: Appl. Phys.* **23** (1990) 1582.

Analysis of optical thickness determination of materials by THz-TDS

O Sushko, R Dubrovka and R S Donnan

School of Electronic Engineering and Computer Science, Queen Mary University of London, Mile End Road, E1 4NS, London, UK

Email: robert.donnan@eecs.qmul.ac.uk

Abstract. Terahertz time-domain spectrometry (THz TDS) is a sensitive probe of the complex dielectric response of materials. Methods vary for converting time-domain response into final material optical parameters together with estimation of associated uncertainties. Here we point out the importance of using an accurate extraction procedure with particular emphasis on the error introduced by associated inaccuracy in thickness determination of a sample. The Total Variation (TV) method is used to estimate sample thickness to sub-micron accuracy, by constructively using the phenomena of multiple internal reflections ('ringing') within a sample. The applicability and performance of the TV methodology is discussed.

1. Introduction

The THz frequency domain is defined to span from 0.1 to 10 THz [1]. THz TDS is a relatively new tool that emerged mainly due to the work of Auston in the field of photoconductive generation [2] and detection [3] of THz radiation. As a separate instrument for measuring material properties in the THz domain, THz-TDS was first used by Grischkovsky *et al.* [4]. Since then THz TDS has been used for characterization of a wide range of materials namely; semiconductors, biomolecules, polymers and even thin metal films [5, 6]. TDS provides a time-domain response function with the absence of the sample providing the conventional reference, system response. Conversion of time-domain traces into dispersion curves of spectral absorption coefficient and refractive index, calls for signal processing techniques to be used. Many factors compound to make signal processing an involved stage during analysis, the most important being the management of system noise. Uncertainty in sample thickness and position, besides other various sources of random and systematic errors are at issue, too. However there are still no internationally-agreed methods for processing the THz response of a sample. Constructive suggestions are therefore proffered in the use of a transfer function for optical parameter extraction and of sample-thickness determination by the total variation (TV) method [14].

2. Extraction Procedures

Determination of estimates of optical constants generally starts with ratioing the (Fourier-transformed) sample response to the reference. This constitutes an experimental transfer function (TF). A theoretical TF is then constructed, that factors in; signal attenuation and retardation of the probing THz, transmission coefficients at the sample-air interface; and internal, multi-order, reflections of the probe-pulse within the sample – a Fabry-Perot (FP) like 'ringing'. The expression for such a TF in the case of a single-layer sample has been described earlier [7] and is formulated as:

$$\tilde{H}(\nu) = \tilde{t}_{12}(\nu)\tilde{t}_{21}(\nu)exp(-i2\pi\nu d(\tilde{n}(\nu) - n_{air})/c) \cdot \sum_{l=0}^{m}[\tilde{r}^2(\nu)exp(-i4d\pi\nu\tilde{n}(\nu)/c)]^l. \quad (1)$$

$\tilde{r}(\nu) = \frac{\tilde{n}(\nu)-n_{air}}{\tilde{n}(\nu)+n_{air}}$, $\tilde{t}_{12}(\nu) = \frac{2n_{air}}{\tilde{n}(\nu)+n_{air}}$, $\tilde{t}_{21}(\nu) = \frac{2\tilde{n}(\nu)}{\tilde{n}(\nu)+n_{air}}$ are the complex Fresnel coefficients at normal incidence; $\tilde{n}(\nu) = n(\nu) - ik(\nu)$ is the complex refractive index of the sample; m is the order of FP reflections in the time-domain (TD) response and can be derived via noting that: $t_{max} \geq nd(1 + 2m)/c$, with t_{max} being the time-elapse for a scan. Note that this model does not include scattering effects taking place on the sample, whose surfaces are assumed to be plane-parallel and flat at the THz energies concerned. The summation term of (1) accounts for FP ringing in the sample. A number of approximations are often made to simplify (1), namely; having Fresnel coefficients to be only real; ignoring higher order reflections in the sample; and, approximation of the FP term as an infinite sum of geometrical progression: $\left(1 - r^2(\nu)exp(-i4d\pi\nu\tilde{n}(\nu)/c)\right)^{-1}$. All these simplifications introduce additional errors to the final estimates of optical parameters, the last being especially problematic (it is only valid for optically thin samples, when $m \geq 5$).

The structure of (1) necessitates the utilization of non-linear regression algorithm for estimation of the complex refractive index by minimization the difference error between theoretical and experimental TFs. Throughout this paper a Levenberg-Marquardt algorithm is used [8], which combines the advantages of Newtonian and gradient descent algorithms.

3. The TV methodology

Error analysis is an essential part of any THz-TDS measurements. An extensive analysis of both systematic and random sources of errors has been done previously [9]. It provides complete information about the accuracy of extracted properties and helps to track and reduce sources of uncertainty. One of the major sources of uncertainty comes from inaccurate determination of sample thickness [10]. The nature of equation (1) is fortuitous in assisting estimation of optical thickness in a very accurate manner.

Various methods have been employed for optical (also read, 'numerical'), thickness determination. Kruger *et al.* [11] used two measurements of the same material but having different thicknesses. After extracting all possible solutions via TF regression analysis, the two coinciding curves corresponded to the actual thickness. The disadvantage of this method is a necessity to reproduce identical conditions for both measurements. It also has a relatively high computational cost. The approach has been developed for determination of thickness for sub-100-μm samples [12]. Its algorithm relies on an additional Fourier transform of the frequency-dependent material parameters to a quasi-space domain. Other methods [13] are based on time-domain reconstruction of the THz response, but they too suffer from high computational requirements. Here a TV method is adopted as a simple and reliable approach for numerical thickness determination [14]. The principle of TV is based on the fact that in most cases the dispersive curves of material properties are smooth, so that an incorrect determination of thickness manifests as spurious oscillations. By iteratively varying thickness, the algorithm selects the best-behaved curve, and this corresponds to the most accurate thickness estimate. The value of TV is calculated for each thickness as:

$$TV = \sum_{f_m}^{f_l}(|n_k - n_{k-1}| + |k_k - k_{k-1}|), \quad (2)$$

where f_m to f_l defines the frequency domain where material parameter curves are smoothest (this typically excludes lower and higher limits of the operating frequency domain). Mathematically, a plot of TV shows the degree of fluctuation in the curves of refractive index and extinction coefficient for each separate thickness value.

Figure 1 shows an example of depth determination for a 1 mm-thick silicon plate. The TV plot clearly shows the minimum that corresponds to the actual value of thickness (the smoothest curve). Figure 1B is a high resolution about the inflection in figure 1A. It estimates the actual thickness of the plate to be 1.0319(1) mm. This is one order of accuracy above the ability of standard micrometer gauges. Multiple micrometer gauge measurements give the thickness of the Si plate to be 1.031±0.001 mm. Note that the sample thickness found numerically corresponds to the effective thickness (taking into account surface roughness and non-uniformity), of the material in the area where the THz beam illumines the sample.

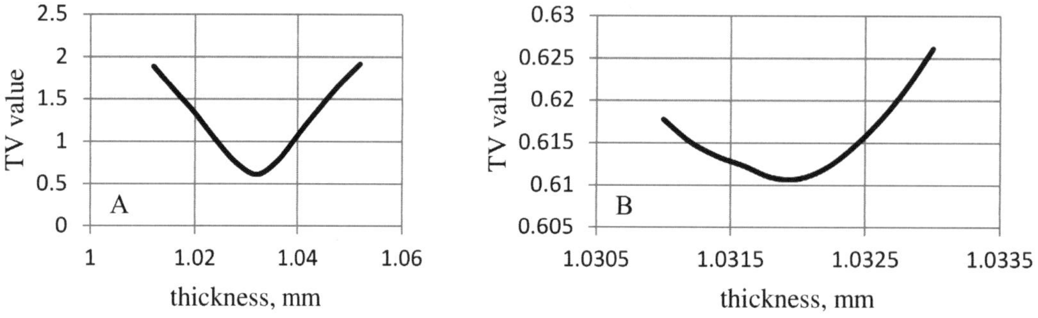

Figure 1. Total variation of material parameter curves, plotted for coarse (A) and denser (B) thickness ranges. The data is shown for a silicon plate of approximately 1 mm thickness.

Successful application of the TV method requires two important conditions to be met: FP-like ringing needs exist and to be above the noise floor of the system. In fact, ringing of the THz pulse within the sample provides the physical means of accurate optical determination of the sample depth. If no internal reflections are detectable in the sample response, the TV algorithm does not work efficiently. In this case, instead of a sharp minimum in the TV curve, a broad, flat minimum extends for 10s of μm. Examination of equation (1) underscores this, where the FP term collapses to unity if no internal reflection is present. Spurious oscillations are consequently absent in parameter dispersion curves. The absence of optical ringing in the sample response can be for several reasons. In measurements of optically-thick samples employing a relatively short scan-length ($t_{scan} < 3nd/c$), when n^{th} order transmissions allied to n^{th} order back-face reflections, simply arrive at the detector after scanning has finished. Interestingly, high absorption by a sample (e.g. $LiNbO_3$), does not lead to a significant decrease in the amplitude of FP-reflections. The feature most responsible for attenuating n^{th} order internal reflection amplitude is scattering of THz radiation both within and at the surfaces of the sample. It is principally this feature that limits the application of the TV methodology to study of bio-material samples like proteins, amino acids, sugars, etc. We have conducted a series of measurements of lysozyme, glycine and lactose pellets; and for all these samples there is no detectable ringing, while system dynamic range remains sufficiently high to distinguish THz pulse ringing within a 0.5 mm thick high-absorbing plate of lithium niobate.

Another significant factor influencing the performance of the TV method is sample alignment. The ideal case (as per the assumption of normal incidence in the use of Fresnel coefficients in (1)), is to set the sample-face-normal in-line with the THz beam axis. But this is not always possible, so a series of measurements have been performed to investigate the effect of face-normal deviation. Sample-tilt corresponds to an effective (secant) increase in sample thickness.

$$d_{new} = d_{actual}/\cos(\alpha) \qquad (3)$$

Figure 2 shows the results of thickness determination of the tilted sample by the TV method and that calculated by (3). Sample thickness calculated by the TV method closely follows prediction (within the angle determination inaccuracy), proving the capability of the method to extract thickness correctly. For a sample tilt of 1° the sample depth changes by 0.3 μm as determined by both TV and equation 3. According to this analysis, in order to ensure thickness-error to be less than 0.1 μm due to sample misalignment, the sample normal needs to be set to within ±0.6°deviation off from beam-axis.

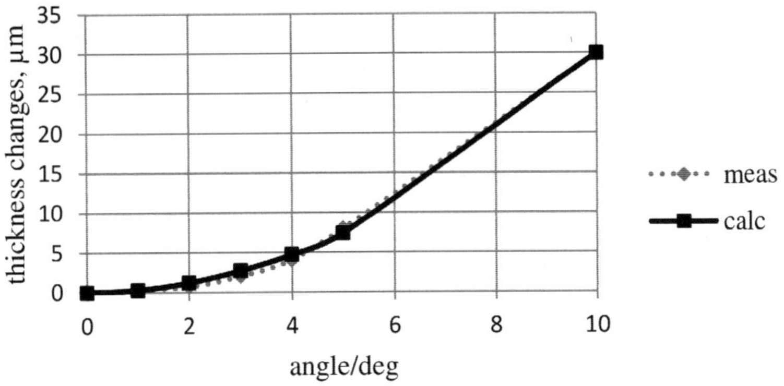

Figure 2. Changes in silicon plate thickness determined from measurements using the TV method, and from equation (3).

4. Conclusions

The governing requirements have been discussed for accurate estimation of material optical properties by THz TDS. The applicability and performance of numerical thickness determination using total variation (TV) technique is assessed. Accurate information on sample-depth, being the major source of uncertainty, is necessary for reliable material parameter (i.e. optical constants) determination. The accuracy of the TV method is shown to reach ±0.1 μm if FP-like multiple-order internal reflection response is present in the sample and alignment of the sample-normal is within ±0.6°of the beam-axis.

References

[1] Fitch M J and Osiander R 2004 *Johns Hopkins APL Tech. Digest* **25** 348-355
[2] Auston D H 1975 *Appl. Phys. Lett.* **26** 101-103.
[3] Auston D H and Smith P R 1983 *Appl. Phys. Lett.* **43** 631-633.
[4] Grischkowsky D, Keiding S, Exter M V and Fattinger C 1990 *J. Opt. Soc. Am.* B **7** 2006-15.
[5] Dai J, Zhang J, Zhang W and Grischkowsky D 2004 *J. Opt. Soc. Am.* B **21** 1379-86.
[6] Zeitler J A, Taday P F, Newnham D A, Pepper M, Gordon K C and Rades T 2007 *Journal of Pharmacy and Pharmacology* **59** 209-223.
[7] Pupeza I, Wilk R and Koch M 2007 *Opt. Express* **15** 4335-50.
[8] More J J 1978 *Numerical Analysis. Lecture Notes in Mathematics* **630** 105-116.
[9] Sushko O, Shala K, Dubrovka R and Donnan R 2013 *J. Opt. Soc. Am.* A **30** 5 979-986.
[10] Duvillaret L, Garet F and Coutaz J-L 1999 *Appl. Opt* **38** 409-415.
[11] Krüger M, Funkner S, Bründermann E and Havenith M 2011 *J. Infrared Mili Terahz Waves* **32** 699-715.
[12] Scheller M, Jancen C and Koch M 2009 *Opt. Comm.* **282** 1304-06.
[13] Scheller M and Koch M 2009 *J. Infrared Mili Terahz Waves* **30** 762-769.
[14] Dorney T D, Baraniuk R G and Mittleman D M 2001 *J. Opt. Soc. Am.* A **18** 1562-71.

Carbon at the Ti site in strontium titanate from first principles

Meaad S AL-Hadidi, J P Goss, P R Briddon, Raied A Al-Hamadany and Mariam E Ahmed

Electrical and Electronic Engineering, Newcastle University, Newcastle upon Tyne, UK

E-mail: m.al-hadidi@ncl.ac.uk

Abstract. Perovskite oxides have attracted both experimental and theoretical attention due to their high permittivities, with $SrTiO_3$ used in waveguides, high capacity computer memory cells and dielectrics. Different growth methods yield various qualities of material, and thin-films growth using organic precursors makes carbon contamination highly probable. Using density functional theory we have calculated the electronic structure and vibrational properties of carbon substituting for Ti in cubic $SrTiO_3$. We find a relatively modest impact upon the electronic structure, with predicted impact upon observable factors, particularly the vibrational spectra and dielectric constant.

1. Introduction

ABO_3 perovskites have been widely studied for use in optoelectronics[1], ferroelectric[2, 3, 4], waveguides[1], high capacity computer memory cells[5, 1] and dielectrics[1]. Indeed, a major challenge in the semiconductor industry is to find suitable high-κ dielectrics to replace the SiO_2 in micro- and nano-electronic devices. To this end, materials including TiO_2, HfO_2 and the perovskite titanates $SrTiO_3$ and $BaTiO_3$ have been highlighted[6, 7]. Bulk crystalline $SrTiO_3$ has a relative permittivity of hundreds at room temperature, making it highly suitable for nano-size devices[8]. However, in thin films leakage current is a major issue, with native point defects, crystallinity and impurities remaining a significant problem.

Different deposition methods have been used to grow $SrTiO_3$ thin films, including chemical vapour deposition[9], molecular beam epitaxy[10] and pulsed laser deposition[11]. Recently, atomic layer deposition, sometimes referred to as atomic layer epitaxy, has attracted more attention due to its accurate thickness control and superior conformal growth[12, 13]. However, the organic Sr precursors lead to contamination[14, 15], and of particular note, carbon impurities have been reported in SrO and $SrTiO_3$ films[16]. In the case of SrO films, the current understanding of carbon contamination[17] is the formation of $SrCO_3$ within the SrO, and the appearance of the carbon contamination may be viewed as an incompleteness of the growth reactions[14].

There are many first principles simulation studies focusing on the electronic properties of impurities in $SrTiO_3$, but relatively few of them relate to carbon. In one density functional simulation[18], carbon is assumed to lie on an oxygen site, where it was found to generate a mid-gap 2p-atomic orbital like, partially occupied state. Another study compared different sites, and suggested that the Ti site is more favourable[19], where it was proposed to introduce

Journal of Physics: Conference Series **472** (2013) 012006 doi:10.1088/1742-6596/472/1/012006

an empty band a little below the host conduction band minimum. However, the comparison in this study examines formation energies where the growth conditions differ with the substitution site, so it is difficult to draw an absolute conclusion from this data.

In this paper, we present the results of first principles calculations on the geometry, electronic structure, vibrational modes and reorientation barrier of carbon substituting for Ti in $SrTiO_3$.

2. Computational details

The properties of carbon substituting for Ti in $SrTiO_3$ are calculated using first-principles density functional theory within the local density approximation[20], as implement in the AIMPRO computer code[21, 22].

Atoms are modelled using norm-conserving, separable pseudo potentials[23] (the valance sets of Sr, Ti, O and C are $4s^24p^65s^2$, $3s^23p^63d^24s^2$, $2s^22p^4$ and $2s^22p^2$, respectively) and the Kohn-Sham eigen-functions are expanded using atom centred Gaussian basis sets[24]. The bases consist of independent sets of s-, p- and d-type Cartesian Gaussian functions of four widths yielding 40 functions per atom for Ti, O and C, while five width of s- and p-Gaussians, yielding 20 functions per atom, centred at Sr. Matrix elements of the Hamiltonian are determined using a plane-wave expansion of the density and Kohn-Sham potential[25] with a cut-off of 150 Ha, resulting in absolute convergence of the total energy with respect to the expansion of the charge density to better than 10 meV.

Structures are optimised using a conjugate gradients scheme, with the optimised structures having forces of $< 10^{-3}$ atomic units and the final structural optimisation step is required to result in a reduction in the total energy of less than 10^{-5} Ha. Although $SrTiO_3$ has an ideal cubic perovskite type structure at room temperature, with the Ti^{+4} ions being six fold coordinated to O^{-2} ions, and each Sr^{+2} surrounded by four TiO_6 octahedron, at very low temperatures, $SrTiO_3$ undergoes[26] a small antiferroelectric distortion with alternate TiO_6 groups rotating about a [001] axis in opposite senses by 1.89°. The low-temperature structure is reproduced in the current computational approach, both in terms of the angle of distortion and the associated energy[27]. For bulk cubic $SrTiO_3$, our computational approach yields an equilibrium lattice parameter of 3.87 Å, within 1% of experiment[28]. The calculated band-gap is 1.85 eV, reflecting the well documented underestimate arising from the underpinning methodology. Our estimate is consistent with previous comparable calculations[29, 30].

To model the defect, we use the standard supercell approximation. In particular, a non-primitive supercell with lattice vectors $2[a\bar{a}\bar{a}]$, $2[\bar{a}a\bar{a}]$ and $2[\bar{a}\bar{a}a]$ is used, where a is the lattice constant in the cubic unit cell. This supercell is comprised from 32 formula units, meaning the supercell is $Sr_{32}Ti_{32}O_{96}$. In all cases the Brillouin zone is sampled using uniform $2 \times 2 \times 2$ Monkhorst-Pack mesh[31].

Vibrational modes have been calculated by obtaining the second derivatives of the energy with respect to atomic positions, which are then assembled into the dynamical matrix. The second derivatives are obtained from a finite difference approximation involving the forces calculated on atom i when displacing atom j in each of the three Cartesian directions. As such, they include even anharmonic terms in the vibrational potential, and the resulting vibrational modes are termed quasi-harmonic. From the dynamical matrix the eigen values yield the squares of the vibrational frequencies and eigen vectors the normal coordinates.

Finally, for activation energies the climbing nudged elastic band method has been used[32, 33], and the convergence of the saddle point energy with respect to the number of images and the image-forces has been established to within a few meV.

3. Results and discussion

3.1. Defect geometry

The substitution of titanium by carbon, C_{Ti}, may be anticipated on the grounds of both C and Ti being group-IV elements. We have optimised a configuration where the initial structure has been perturbed form the ideal octahedral symmetry, avoiding any artificial symmetry constraint. We find that the relaxed structure has three C–O bonds significantly shorter (1.29 Å) than the Ti–O inter-nuclear distances in $SrTiO_3$ (1.93 Å). The calculated C–O bond-lengths in C_{Ti} are close to those of the carbonate group[34, 35], so that the relaxation of the structure into the highly distorted geometry can be interpreted in terms of formation of a carbonate ion, CO_3^{-2}, associated with a single Sr^{+2} cation. The structure is shown schematically in figure 1, and after structural optimisation we find it to have very close to C_{3v} symmetry. Here, carbon forms covalent bonds with three of its O neighbours, leaving three of oxygen ions effectively under co-ordinated, but overall the anions and cations are balanced. For the simulation supercell, $(SrTiO_3)_{32}$, one might express the substitution of Ti by C as resulting in $(SrTiO_3)_{31}(SrCO_3)_1$, or approximately $(SrTiO_3)_{0.97}(SrCO_3)_{0.03}$.

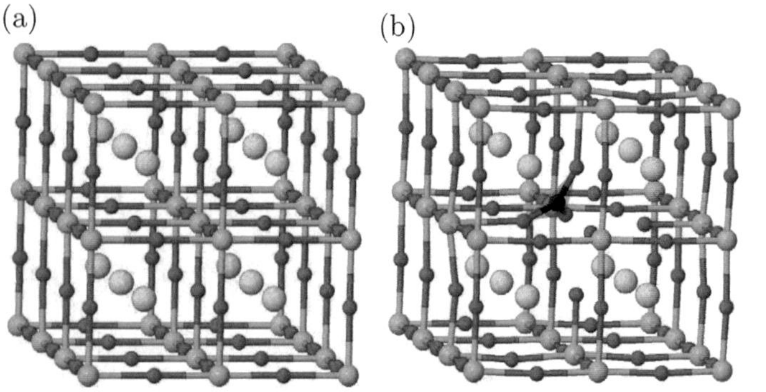

Figure 1. Schematic of (b) C_{Ti} in $SrTiO_3$, with an equivalent section of defect free $SrTiO_3$ shown in (a) for comparison. Green, grey, red and black atoms represent Sr, Ti, O and C respectively. Vertical and horizontal axes are approximately [001] and [010], respectively, with the tilted view adopted to aid clarity.

We now turn to the impact that the formation of the carbonate group has upon the electronic properties of the material.

3.2. Electronic structure

The calculated electron band structure for C_{Ti} superimposed upon the band edges of pure $SrTiO_3$ is shown in figure 2. We first note that the inclusion of carbon has the apparent effect of increasing the band-gap. In fact, the extent of the increase in band gap is found to depend upon the effective concentration of C, with the increase larger where C_{Ti} is modelled in a smaller simulation cell. This is to be expected as the band gap of pure $SrTiO_3$, at 3.25 eV, is smaller than that of $Sr(CO_3)$ at around 4.3 eV[36], so the 1% alloy might well be expected to have a band gap slightly larger than pure $SrTiO_3$.

In addition to the calculation of the band structure, preliminary calculations for the static dielectric function suggest a reduction in the presence of C_{Ti}, consistent with the increase in the band-gap. We find that the electronic part of the static dielectric function[30] is reduced by around 7% for a C:Ti ratio of 1:31.

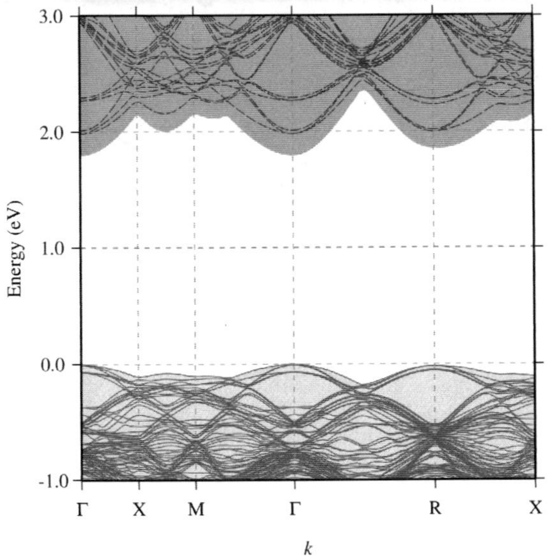

Figure 2. Calculated band structure of C_{Ti} in $SrTiO_3$ in the first Brillouin zone, plotted along high-symmetry branches. Labelling on the x-axis follows the conventions for special points at the zone-boundary for simple cubic Brillouin zone. Occupied and empty bands are shown in solid (green) and dashed (red) lines, respectively. The underlying shaded areas show the regions of the bands for the corresponding C-free supercell. The zero on the energy scale is defined to be the highest occupied state in each case.

3.3. Vibrational modes

To provide a mechanism for direct identification of C_{Ti} in experiment, a local vibrational mode above the one-phonon maximum could prove highly effective. Of particular note, for the C_{3v} C_{Ti} centre we found a degenerate (E) C–O stretch mode at 1414 cm^{-1}, and a breathing stretch-mode with A_1 symmetry at 1081 cm^{-1}. These are close to the values of the D_{3h} carbonate group [37] which have E' and A_1' stretch modes lying around 1415 cm^{-1} and 1064cm, respectively. Lower frequency modes of the carbonate group are also present for C_{Ti}, but are resonant with the one-phonon density of states. Both of the local modes are infra-red and Raman active, and given sufficient concentrations one might seek to confirm the presence of this form of carbon centre directly via either IR or Raman spectroscopy.

3.4. Carbonate group reorientation

Finally, we have examined the energetics for reorientation of the carbonate group. The initial and final structures, and the energy profile along the minimum energy path between them, are shown in figure 3. The activation energy for reorientation is predicted to be around 2.5 eV. This relatively high value is a consequence of the covalent bonding within the carbonate group, so that reorientation involves both the breaking and formation of strong chemical bonds. The carbonate group may be oriented with the principle axis along any of the eight $\langle 111 \rangle$ directions if randomly orientated. However, under specific conditions, such as (111) biaxial strain induced by interface mismatch, the different orientations would have different energies, and given the high energy for reorientation, may be preferentially aligned. Since the C_{Ti} structure has a permanent electric dipole, a polarised population of defect centres that are unable to reorient under normal temperature and electric field conditions, would be expected to have an impact upon the dielectric anisotropy and, where ferroelectric distortions are induced under strain, C_{Ti} would be expected to alter both the details of the phase, and the resulting dielectric constant.

4. Summary and conclusions

In summary, we have studied the geometry, electronic structure and vibration modes for a C impurity substituting for titanium in cubic $SrTiO_3$ using first principles calculations. We find that C_{Ti} represents the formation of carbonate groups, supported by both the geometry of the optimised structure and the calculated vibrational modes for this defect. Furthermore, by

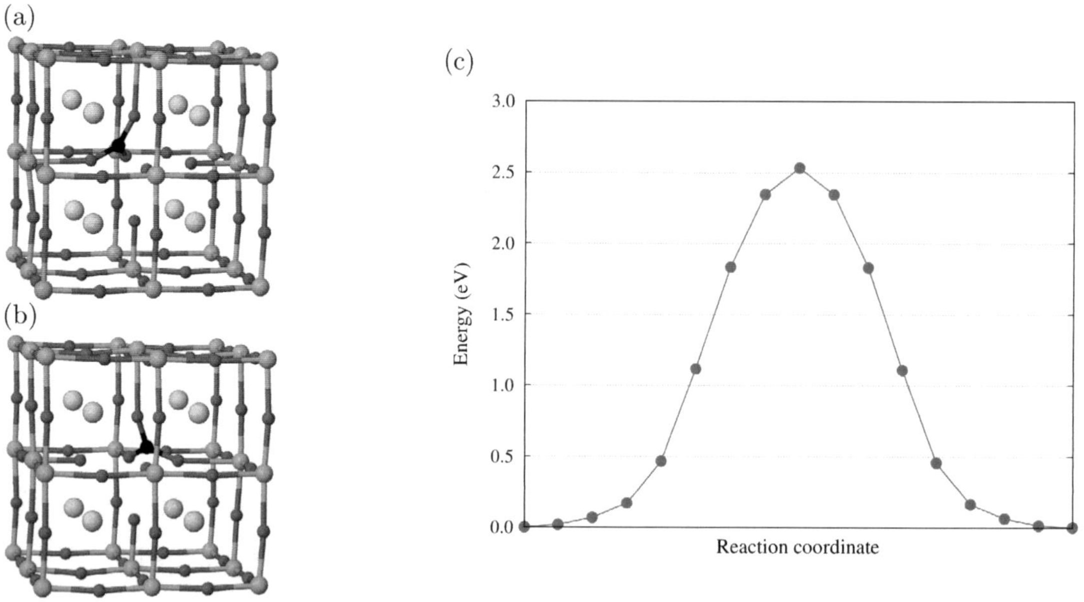

Figure 3. Schematic of (a) initial, and (b) final structure of C_{Ti} in $SrTiO_3$ in a simple reorientation process. Colours and orientation are as indicated in figure 1. (c) Calculated reorientation barrier of C_{Ti} in $SrTiO_3$ between initial and final structure.

comparison between the band structure for pure $SrTiO_3$ and the impurity system, carbon in this form does not directly change the conduction electron or hole concentrations of the system by doping, but by an increase in the band-gap in high carbon concentration materials the thermally generated concentration of carriers is expected to be reduced. Therefore, C_{Ti} is predicted to have a modest impact upon the electrical conductivity of $SrTiO_3$ at high concentrations, and is not responsible for an increase in leakage current in capacitive device elements. Finally, we also estimated the reorientation barrier for the carbonate group, and because the process involves the breaking of C–O covalent bonds, the barrier is a relatively high value at around 2.5 eV. Such a barrier when combined with the fact that C_{Ti} has a permanent electric dipole is anticipated to have an impact upon the dielectric properties of $SrTiO_3$ in an alternating electric field.

References

[1] Pontes FM, Leite ER, Lee JH, Longo E and Varela JA, 2001 *J. Euro. Ceramic Soc.* **21** 419–426
[2] Ahn CH, Rabe KM and Triscone JM, 2004 *Science* **303** 488–491
[3] Kan D, Kanda R, Kanemitsu Y, Shimakawa Y, Takano M, T. T and Ishizumi A, 2006 *Appl. Phys. Lett.* **88** 191916
[4] Jang HW, Kumar A, Denev S, Biegalski MD, Maksymovych P, Bark CW, Nelson CT, Folkman CM, Baek SH, Balke N, Brooks CM, Tenne DA, Schlom DG, Chen LQ, Pan XQ, Kalinin SV, Gopalan V and Eom CB, 2010 *Phys. Rev. Lett.* **104** 197601
[5] Bellingeri E, Pellegrino L, Marré D, Pallecchi I and Siri AS, 2003 *J. Appl. Phys.* **94** 5976–5981
[6] Heh D, Young CD, A. BG, Hung PY, Diebold A, Bersuker G, M. VE and Bernstein JBB, 2006 *Appl. Phys. Lett.* **88** 152907
[7] Chevallier J, Clerjaud B and Pajot B, 1991 *Neutralization Of Defects And Dopants In III-V Semiconductors*, volume 34 of *Semiconductors and Semimetals*, chapter 13. Academic Press, San-Diego, pages 447–510
[8] Choudhury BK, Rao KV and Choudhury RNP, 1989 *J. Mol. Struct.* **24** 3469–3474
[9] Song Z, Rogers BR and Theodore ND, 2004 *J. Vac. Sci. and Technol. A* **22** 711–718
[10] McKee RA, Walker FJ and Chisholm MF, 1998 *Phys. Rev. Lett.* **81** 3014–3017
[11] Yan L, Suchomel MR, Grygiel C, Niu HJ, McMitchell SRC, Bacsa J, Clark JH, Werner M, Chalker PR and Rosseinsky MJ, 2009 *Appl. Phys. Lett.* **94** 232903

[12] Puurunen RL, 2005 *J. Appl. Phys.* **97** 121301
[13] Zhang CB, Wielunski L and Willis BG, 2011 *Appl. Surf. Sci.* **257** 4826–4830
[14] Kosola A, Putkonen M, Johansson L and Niinistö L, 2003 *Appl. Surf. Sci.* **211** 102 – 112
[15] Willis BG and Mathew A, 2008 *J. Vac. Sci. and Technol. A* **26** 83–89
[16] Lee SW, Han JH, Han S, Lee W, Jang JH, Seo M, Kim SK, Dussarrat C, Gatineau J, Min Y and Hwang CS, 2011 *Chem. of Materials* **23** 2227
[17] Vehkamäki M, Hänninen T, Ritala M, Leskelä M, Sajavaara T, Rauhala E and Keinonen J, 2001 *Chem. Vapor Dep.* **7** 75–80
[18] Liu HF, 2012 *Solid State Commun.* **152** 2063–2065
[19] Li N and Yao KL, 2012 *AIP Advances* **2** 032135
[20] Perdew JP and Wang Y, 1992 *Phys. Rev. B* **45** 13244–13249
[21] Briddon PR and Jones R, 2000 *Phys. Status Solidi B* **217** 131–171
[22] Rayson MJ and Briddon PR, 2008 *Computer Phys. Comm.* **178** 128–134
[23] Hartwigsen C, Goedecker S and Hutter J, 1998 *Phys. Rev. B* **58** 3641–3662
[24] Goss JP, Shaw MJ and Briddon PR, 2007 In *Theory of Defects in Semiconductors*, edited by DA Drabold and SK Estreicher, volume 104 of *Topics in Applied Physics*. Springer, Berlin/Heidelberg, pages 69–94
[25] Rayson MJ and Briddon PR, 2009 *Phys. Rev. B* **80** 205104
[26] Lytle FW, 1964 *J. Appl. Phys.* **35** 2212
[27] Al-Hamadany R, Goss JP, Briddon PR, Mojarad SA, Al-Hadidi M, O'Neill AG and Rayson MJ, 2013 *J. Appl. Phys.* **113** 024108
[28] Evarestov RA, Piskunov S, Kotomin EA and Borstel G, 2003 *Phys. Rev. B* **67** 64101
[29] Samantaray CB, Sim H and Hwang H, 2005 *Microelec. Journal* **36** 725
[30] Marques M, Teles LK, Anjos V, Scolfaro LMR, Leite JR, Freire VN, Farias GA and da Silva EF, 2003 *Appl. Phys. Lett.* **82** 3074
[31] Monkhorst HJ and Pack JD, 1976 *Phys. Rev. B* **13** 5188–5192
[32] Henkelman G, Uberuaga BP and Jónsson H, 2000 *J. Chem. Phys.* **113** 9901–9904
[33] Henkelman G and Jónsson H, 2000 *J. Chem. Phys.* **113** 9978–9985
[34] Adler HH and Kerr PF, 1963 *Am. Miner.* **48** 839–853
[35] Gagliardi L and Roos BO, 2002 *Inorg. Chem.* **41** 1315
[36] Ci Z and Wang Y, 2009 *J. Electrochemical Soc.* **156** J267–J272
[37] Tavender SM, Johnson SA, Balsom D, Parker wA and Bisby RH, 1999 *Laser Chem.* **19** 311–316

Bridging in contaminated transformer oil under DC and AC electric field.

Shekhar Mahmud*[1], George Chen[1], Igor O. Golosnoy[1], Gordon Wilson[2] and Paul Jarman[2]
[1]University of Southampton, Southampton, SO17 1BJ, UK
[2]National Grid, UK

*E-mail: sm8e08@ecs.soton.ac.uk

Abstract. There are two experimental set up have been established to investigate the bridging effect between two electrodes with different potentials under ac and dc voltages. This paper will explain the full details of the experimental setup and the results. These experiments carried out on pressboard particles under three different levels of ac and dc voltages. Several contamination levels have been investigated. Optical images of particle accumulation have been recorded along with associated conduction current measurement during experiments. At higher voltages the rate of bridge formation is increased along with an associated current increase. The major differences between ac and dc bridges are the shapes of the particle accumulation. The bridge formation rate for ac is slower than dc.

1. Introduction

One of the key components in high voltage transmission and distribution systems are power transformers and their consistent operation is of utmost importance to energy utility companies and end users. However, transformer failures in the systems have gathered pace as more transformers are approaching or exceeding their designed lifetime. Almost 30% of total transformer failures are caused by insulation/oil contamination which has been revealed by recent analysis [1]. Therefore, it is essential to understand the failure mechanisms/processes so the proper measures can be taken to prevent transformer failure and this way we can save the huge cost caused by it in terms of asset loss and business interruption.
Liquid dielectrics are used as insulation and cooling agent in almost all the high voltage power transformers. This petroleum based dielectric liquid is called “transformer oil”. One of the major disadvantages is the transformer oil can be easily contaminated [2, 3]. During packaging, shifting, storing the transformer oil can be contaminated. During the operation it contacts with metal, iron core and pressboard insulation inside a transformer. Contaminants such as metal filings or cellulosic residual can also be formed in the oil, especially for transformers with aged paper insulation. During normal operation non-uniform fields are present within the transformer. These contaminants tend to move towards high field regions due to dielectrophoresis (DEP) forces and could form a bridge over a period of time. The dielectrophoresis phenomenon first introduced by Phol [4] whereby a neutral body placed in an electric field becomes polarized and is equivalent to an electric dipole. One side of the dipole has an excess of positive charge and negative charge on the other. Under non-uniform electric field region, the forces acting on the two ends do not balance and the particles move towards either

high or low field region. The bridge may potentially act as a conducting path between two different potentials within the transformer structure, leading to partial discharges or insulation failure.

Many researchers have studies various aspects of bridging phenomenon in transformer oil, such as, conduction current, partial discharge, resistivity [5, 6, and 7] etc. Our previous studies were focused on the effect of particle size under DC electric field [8] and numerical simulation of the bridge in transformer oil [9]. This paper has particularly concentrated on the different mechanism of particle accumulation under the influence of DC and AC electric field.

The aim of current work is to distinguish the particle accumulation process in contaminated transformer oil under the influence of AC and DC high electric field. Two different experiments have been set up to investigate the bridging effect between two electrodes with different potentials. The experimental setup and the results have been discussed in subsequent sections. These experiments have been carried out with pressboard particles under the influence of three different DC (2kV, 7.5kV and 15kV) and AC (10kV, 15kV and 20kV) voltages. There are several contamination levels of cellulose particles have been investigated ranging from 0.001% to 0.024% by weight. Optical microscopic images and conduction current of the bridge formation have been recorded for all the experiments.

2. Experiments

2.1. Sample tank

We have used a glass built sample tank for all the experiments. The internal volume of the tank is 550 ml. There were two spherical electrodes with the diameter of 13mm placed in the middle of the cell and they were attached to side walls the of the test cell. The electrodes gap is adjustable via a screw and it was kept constant at 10 mm for all the experiments.

2.2. Preparations of samples

A new pressboard which normally used in high voltage transformers was rubbed with metal files to produce the fiber like dust particles. The fibers were categorised in to different sizes using sieves. Four different sizes of particles were produced, less than 63 μm, 63-150 μm, 150-250 μm and 250-500 μm. The particles were separated by the fiber width rather than length. We have tested only 150-250 μm particles under the influence of AC electric field. The contamination levels for the particles were 0.001%, 0.002%, 0.003%, 0.006%, 0.008%, 0.016% and 0.024% by weight.

We have cleaned the sample tank with soap solution in hot water and dried under hot air flow before starting any test with new particle size. The tank was first rinsed with some clean oil to repeat an experiment with same size particles and wiped thoroughly with cyclohexene.

We have filled test cell with 300 ml of Gemini X type mineral transformer oil from Nynas to start a new experiment. The quantity of transformer oil was enough to submerge the electrodes completely. The contaminants of lowest level was then added to the oil. The sample tank was covered with cling film to protect from external dust and moisture and it remained covered for the whole period of experiment apart from adding the next level of contaminants. The sample tank was stirred for 2 minutes prior to every test on a magnetic stirrer which enable us to distribute the particles evenly thought the transformer oil.

2.3. Experimental Setup

To record the optical images of the bridging, the test cell was placed under a digital camera mounted stereo microscope. We have positioned the microscope inside an aluminum box which acts as a Faraday Cage. The camera, Keithley Picoammeter 6485A and multimeter 2001 were controlled from a computer and the data were also saved on that computer. The high voltage supply were calibrated at the beginning of the experiments and an external circuitry used to monitor the voltage to ensure supply voltage is same for every test. Block diagrams of complete experimental setup is shown on figure 1 and 2.

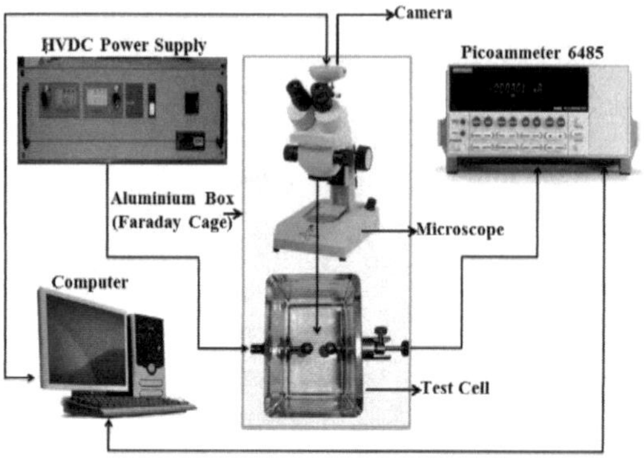

Figure 1. Complete block diagram of experimental setup for DC test [7]

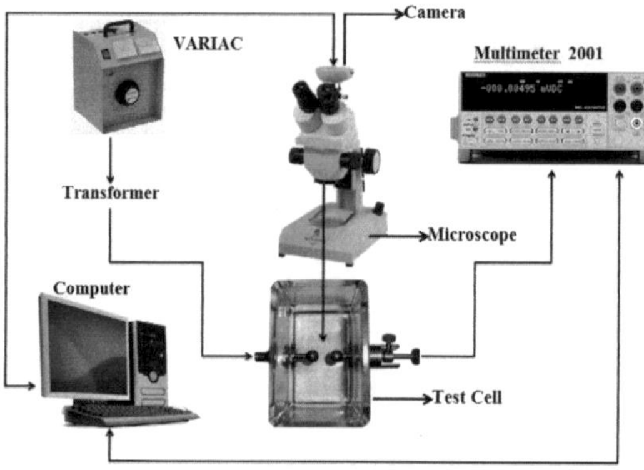

Figure 2. Complete block diagram of experimental setup for AC test

2.4. Testing procedure

One of the spherical electrodes was attached to the high voltage power supply and other electrode was connected to the ground via a Keithley picoammeter 6485 for DC and multimeter 2001 for AC to measure the conduction current through the gaps. Three different levels of DC (2kV, 7.5kV and 15kV) and AC (10kV, 15kV and 20kV) high voltages were applied to the samples. The experiments were conducted until a complete bridge was created or maximum of 25 minutes where there was no bridge. The digital camera was set to take images in a regular interval during the test to record the images of particle accumulation. The tests were carried out at room temperatures. The temperatures were kept within 3 degrees difference for all the DC experiments but the temperatures were not controlled for AC experiments. Each test was carried out three times for a particular voltage level to observe the repeatability.

3. Results and Discussion

3.1. Bridge formation with 0.001% contaminants

The particles were moving back and forward between the electrodes as soon as 2kV DC was applied. They were charging from one electrode and discharging at the other. The pressboard fibers were started attaching on the electrode surfaces. Longer fibers was attached to the electrodes and then they elongated towards the other electrode. Small spherical particles started attaching themselves to the long fibers afterwards. A complete bridge continued to grow in this way. A very thin bridge formed for 2kV supply after 600s for on figure 3. Applied voltage of 7.5 kV initiated the particles movement more quickly. Within only 240s A thin bridge were formed. The bridge continued to grow thicker until 600s (figure 3). No change was observed after that. The completed bridge for the bigger particles was not as thick as for the smaller particles. The particles started to move fiercely after applying 15kV DC. A thin complete bridge was formed by 70s and it continued to thicken up to 300s. The bridges were thicker near the electrodes and thinner towards the middle.

The particles were slowly moving after the 10kV AC supply turned on. They started to move towards the high gradient of electric field region and started attaching themselves to the electrodes slowly because positive of dielectrophoretic force. This process of attaching to the electrodes were going on equally to both electrodes. More particles were attached to the electrodes as time elapsed. These experiments were carried out on several contamination levels i.e. 0.001%, 0.002, 0.003, 0.006%, 0.008%, 0.016% and 0.024%. As the concentration increased the particle accumulations to the electrodes increased. But they never formed a complete bridge between the two electrodes. Our experimental result does not align with [6], in which bridging were observed with spherical and wrapped electrode system in moistened and pressboard fiber concentrated samples under AC electric field under the influence of 6kV and 8 kV AC electric field for spherical and formed wire electrodes respectively.

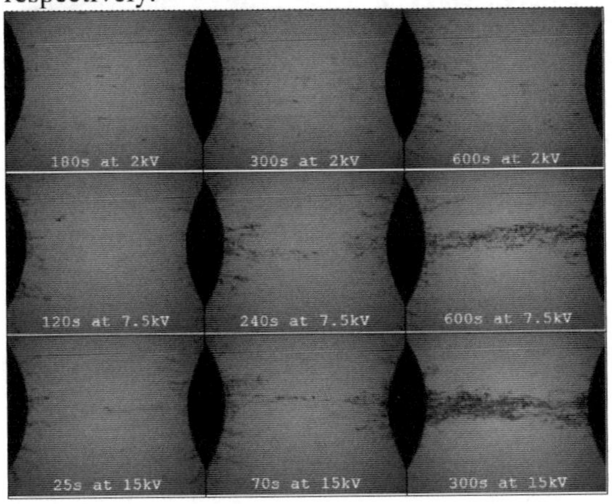

Figure 3. Optical microscopic images of bridging in contaminated transformer oil with 150-250 μm pressboard fiber, concentration level 0.001% under DC [8]

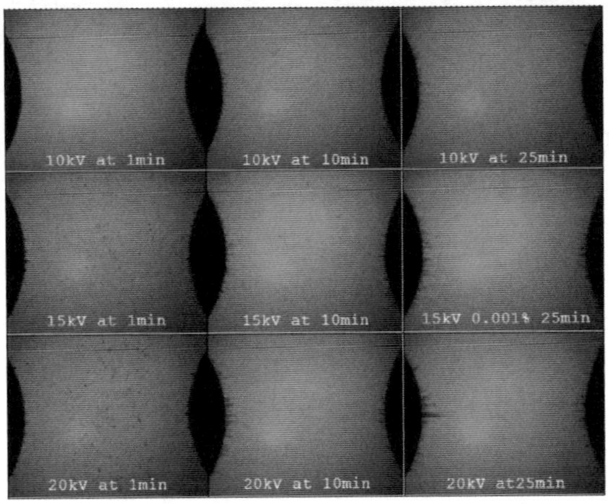

Figure 4. Optical microscopic images of bridging in contaminated transformer oil with 150-250 μm pressboard fiber, concentration level 0.001% under AC

3.2. Bridge formation and conduction current with 0.003% contaminants for DC and 0.024% for AC

The contaminants were started moving slowly between the electrodes after applying 2kV DC electric field. This movement of particles was due to charging and discharging. A complete thin bridge was observed after 180s. The bridge grew until 600s for as shown on figure 5. Different branches were made from one electrode to the other. It took only 10s to form a bridge after 7.5kV applied. The bridge was thickened until 300s, no change observed after that. The completed bridge for lower voltage was shallower than higher voltage. The movements of the particle were intensified after applying 15kV. A thin complete bridge was made within 5s, and thickened up to 60s.

The movement of the particles were intensified with the applied voltage increased from 10kV, 15kV to 20kV AC. The particle accumulations towards the high electric field region were also increased with the applied voltage. This test with 0.024% contamination had the highest particle attachment to the electrodes. But no complete bridges were formed between the electrodes.

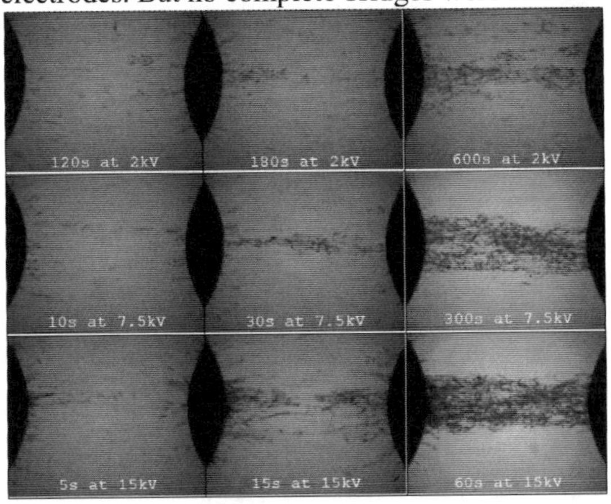

Figure 5. Microscopic images of bridging in contaminated transformer oil, concentration level 0.003% under DC [8]

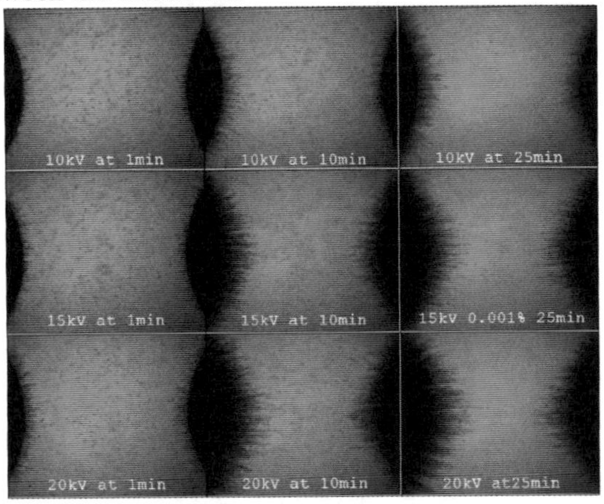

Figure 6. Microscopic images of bridging in contaminated transformer oil, concentration level 0.024% under AC

Conduction currents with contamination levels of 0.001% to 0.004% and 0.001% to 0.024% shown in the figure 7 and 8 under the influence of 15kV DC and 15kV AC respectively. From the DC current plot we can observe that the current increased non-linearly with the increment of contamination level under a fixed voltage. When the contamination level linearly increased from 0.001% to 0.004%, the conduction current increased from 3, 6, 15, and 23 nano amp which means 2, 5 and 8 times with respect to lowest contamination level of 0.001%. There was no noticeable change exist in the AC conduction current. The contamination levels were tested from 0.001% up to 0.024% but as there was no complete bridge formed, the current also didn't conducted through the particles.

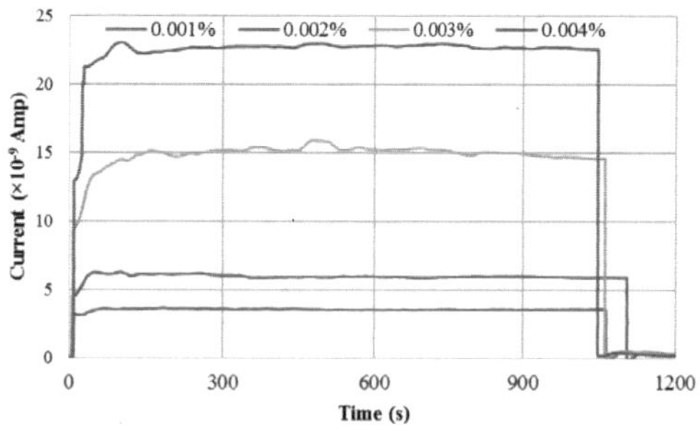

Figure 7. Conduction current with different contamination levels under the influence of 15kV DC electric field.

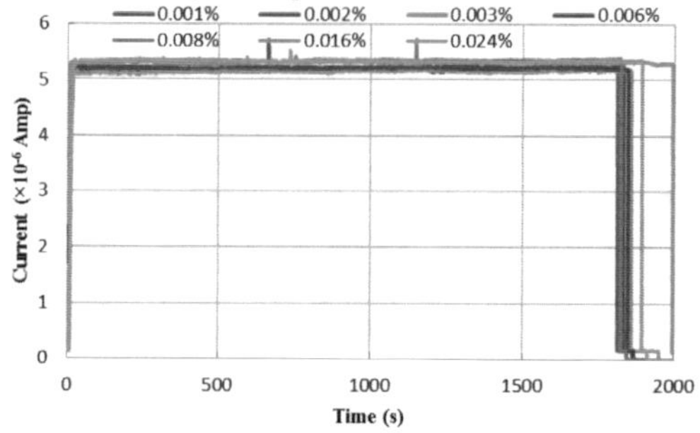

Figure 8. Conduction current with different contamination levels under the influence of 15kV AC electric field.

3.3. Comparison between AC and DC conduction current

The conduction currents of all the contamination levels are summarised in figure 9 for DC and figure 10 for AC. The maximum currents were plotted where the values almost steady. The DC conduction currents were almost linear with applied voltage of 7.5kV. Though some non-linearity features of the process can be clearly seen for 15kV but overall we can say that the current increased with the increment of contamination levels for DC case. Two different mechanism can contribute to the conduction, namely charging of separate particles by touching the electrodes together with their migration and formation of the conductive bridge. On the other hand, the conduction currents for AC were almost constant for all the concentration levels tested. There might be various reasons, there was no complete bridge formed between the electrodes, the particles were not charging because of alternating frequency quicker than charging time. It was also reported that the overall conductivity of the particles and the medium affects under low frequency but the overall permittivity affects by high frequency.

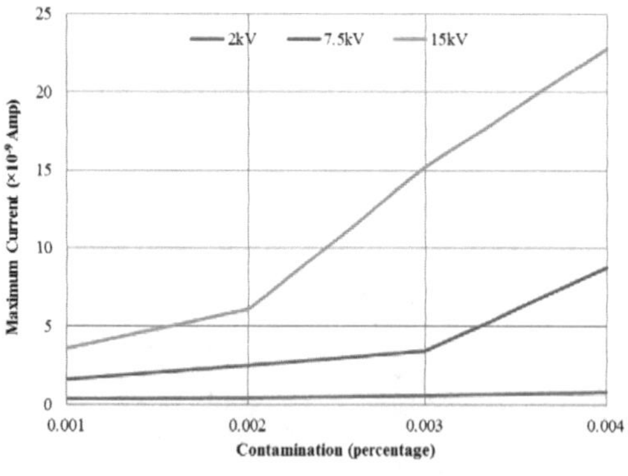

Figure 9. Conduction current under DC.

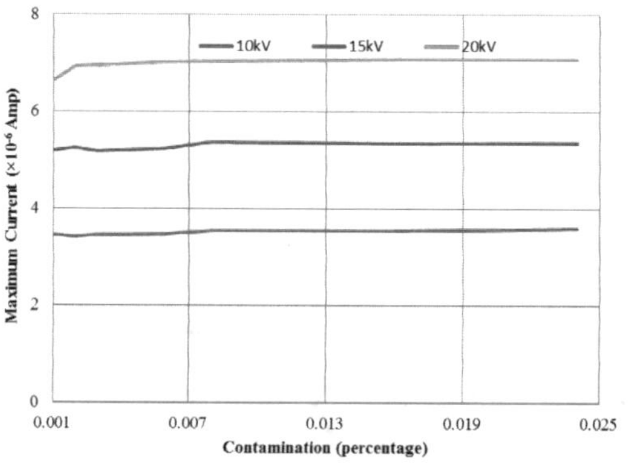

Figure 10. Conduction current under AC

4. Conclusion

The major difference in particle accumulation under the influence of DC and AC voltage is the shape of the particles assembly. It can be concluded that at higher DC voltages, the rate of bridge formation is strongly increased along with an associated current increase. It takes longer time to accumulate the particles AC electric field compared to DC. The particles are strongly bonded, covered the half of the electrode surface and slow increase in the middle of the electrode for the AC case. A complete bridge between the two electrodes is always formed no matter how small is particle concentration level in transformer oil whereas for AC electric field there was never a case of full bridge formation between the electrodes even with the concentration level 24 time higher than the DC electric field. Future studies will include the experiments under DC biased AC electric field, effect of electrode shape and tests with insulated electrodes to comprehend the dynamics of particle accumulation in transformer oil.

Acknowledgements
Authors would like to thank National Grid and IET Power Academy for sponsoring the project.

References

[1] Sokolov, V. V., 1994. Experience with the refurbishment and life extension of large power transformers. *Minutes of the sixty-first annual conference of doble clients,* Sec. 6-4.
[2] Rouse, T. O., 1998. Mineral insulating oil in transformers. *Electrical Insulation Magazine, IEEE* 14(3): 6-16.
[3] Danikas, M.G., 1990, Breakdown of transformer oil. *Electrical Insulation Magazine, IEEE.* 6(5): p. 27-34.
[4] Pohl H. A., 1951, The motion and precipitation of suspensoids in divergent electric fields. *Journal of Applied Physics*. 22(7): p. 869-871.
[5] Chen G. and Zuber M. H., 2007, Pre-breakdown characteristics of contaminated power transformer oil, *Electrical Insulation and Dielectric Phenomena*, CEIDP 2007. Annual Report - Conference on, 2007, pp. 659-662.
[6] K. W. H. Moranda, H. M. Grzesiak, 2003, Dynamics of bridge creating in contaminated oil at AC voltage and analysis of accompanying partial discharges, *XIII International Symposium on High Voltage Engineering*, Netherlands.
[7] J. G. M. Ossowski, H. Moronda and H. Moscicka-Grzesiak, 2003, Oil resistance at different phases of bridge mechanism development at direct voltage, *XIII International Symposium on High Voltage Engineering,* Netherlands.
[8] Mahmud, S., Chen, G., Golosnoy, I. O., Wilson, G. and Jarman, P., 2012, Bridging phenomenon in contaminated transformer oil. *Proceedings of 2012 International Conference on Condition Monitoring and Diagnosis. 2012 International Conference on Condition Monitoring and Diagnosis.* Piscataway, US, Institute of Electrical and Electronics Engineers, 180-183.
[9] Mahmud, S., Golosnoy, I. O., Chen, G., Wilson, G. and Jarman, P., 2012, Numerical simulations of bridging phenomena in contaminated transformer oil. *Proceedings of 2012 IEEE Conference on Electrical Insulation and Dielectric Phenomena,* Montreal, Canada, 14 - 17 Oct 2012.4pp, 383-386.

Journal of Physics: Conference Series **472** (2013) 012008 doi:10.1088/1742-6596/472/1/012008

Permittivity measurement of minerals

VJ Anderson, A Clarke, GH Meeten
Schlumberger Gould Research, Madingley Road, Cambridge CB3 0EL UK

Email: meeten1@slb.com

Abstract. An electromagnetic logging tool moves along a borehole to measure continuously the complex dielectric permittivity of the rock formation. Dielectric properties of formation fluids and minerals are required to interpret the log. Methods to measure mineral permittivities are reviewed and experimental results are described for various particulate minerals.

1. Wellbore logging

Early borehole logging tools measured the kHz electrical resistivity ρ_F of a rock formation of porosity ϕ, with oil and brine in its pores, to obtain the oil volume fraction f_O. Log interpretation used Archie's law [1] $\rho_B/\rho_F \approx f_B^2$ (for brine resistivity ρ_B) to obtain $f_O = \phi - f_B$. This method could discriminate only between brine ($\rho_B \approx 1$ ohm m) and oil ($\rho_O \approx 10^9$ ohm m). Modern electromagnetic well-logging [2] measures the formation permittivity ε and conductivity σ in the MHz – GHz range. Permittivity is material-specific, e.g. oil ($\varepsilon \approx 2$), rock minerals ($\varepsilon \approx 4 - 16$), brine ($\varepsilon \approx 70$). This paper describes a method to measure the permittivity of solid granular minerals required for dielectric log interpretation.

2. Permittivity-matching

Mixtures of dissimilar dielectric materials, such as a liquid-saturated rock or mineral particles suspended in a liquid, have a real permittivity $\varepsilon(\varepsilon_i,q_i)$ that depends largely on the mixture's volumetric composition (q_1, q_2, etc.) and on its dielectric constitution (ε_1, ε_2, etc.).

We compare various formulations of $\varepsilon(\varepsilon_i,q_i)$, particularly for two-component systems where $\varepsilon_1 \approx \varepsilon_2$, and describe the permittivity-matching method to obtain rock mineral permittivity. Subterranean minerals usually are available only in granular form of irregular particle shape so that methods for homogeneous materials cannot be used. Rock formation minerals are largely insulating and we shall assume the permittivity $\varepsilon_{\min}$ to be real. Simple methods measure the permittivity ε_{sus} of a suspension of the mineral with a liquid of known permittivity ε_{liq}, and use a mixture rule that describes ε_{sus} in terms of $\varepsilon_{\min}$, ε_{liq}, and the respective volume fractions [3]. However, such rules are approximate and various rules will give different $\varepsilon_{\min}$ from the same experimental data. Here we describe the dielectric matching method where the particle permittivity is obtained from the liquid permittivity which is unaltered by adding suspended particles. This method is independent of particle shape and concentration, and was used to obtain the porosity and water content of soils, sediments, and rocks [4].

2.1. Theory

General theorems exist for a two-component heterogeneous medium having a structural length-scale large compared with molecular dimensions but small compared with the size of the sample. To define effective dielectric parameters of such microscopically-heterogeneous and statistically-homogeneous media the volume-averaged fluxes and fields in the media (assumed isotropic) can be generally related. The effective permittivity ε is given by $\overline{\boldsymbol{D}} = \varepsilon \overline{\boldsymbol{E}}$ in terms of the volume-average flux and field vectors $\overline{\boldsymbol{D}}$ and $\overline{\boldsymbol{E}}$. Statistical continuum theory [5,6] gives the permittivity ε of a loss-free dielectric as

$$\varepsilon = \bar{\varepsilon} - \overline{(\varepsilon(\boldsymbol{r}) - \bar{\varepsilon})^2} / (3\bar{\varepsilon}) \tag{1}$$

for weak fluctuations of $\varepsilon(\boldsymbol{r})$ about the volume average permittivity $\bar{\varepsilon}$, i.e. $\varepsilon(\boldsymbol{r}) - \bar{\varepsilon} << \bar{\varepsilon}$, as pertains close to permittivity-match in the matching method. In (1) $\varepsilon(\boldsymbol{r})$ describes the spatial fluctuation of the permittivity via vector $\boldsymbol{r}$, and the volume average of $\varepsilon(\boldsymbol{r})$ is

$$\bar{\varepsilon} = V^{-1}\int_V \varepsilon(\boldsymbol{r})\,\mathrm{d}V \tag{2}$$

where $\overline{(\varepsilon(\boldsymbol{r})-\bar{\varepsilon})^2}$ is the volume-average square deviation of $\varepsilon(\boldsymbol{r})$ from $\bar{\varepsilon}$. For an homogeneous dielectric of permittivity ε then $\overline{(\varepsilon(\boldsymbol{r})-\bar{\varepsilon})^2}=0$ and (1) gives the expected result $\varepsilon \equiv \bar{\varepsilon}$. For an heterogeneous dielectric $\overline{(\varepsilon(\boldsymbol{r})-\bar{\varepsilon})^2}>0$ regardless of the permittivity spatial distribution and (1) shows that $\varepsilon < \bar{\varepsilon}$. If a composite dielectric has components #1 and #2 of volume fractions q_1 and q_2, and permittivities ε_1 and ε_2, then $\bar{\varepsilon} = q_1\varepsilon_1 + q_2\varepsilon_2$. As $\overline{(\varepsilon(\boldsymbol{r})-\bar{\varepsilon})^2} = \overline{\varepsilon(\boldsymbol{r})^2} + \bar{\varepsilon}^2 - 2\bar{\varepsilon}\,\overline{\varepsilon(\boldsymbol{r})}$, $\overline{\varepsilon(\boldsymbol{r})} = \bar{\varepsilon}$, and $\overline{\varepsilon(\boldsymbol{r})^2} = q_1\varepsilon_1^2 + q_2\varepsilon_2^2$ then (1) gives

$$\varepsilon = \bar{\varepsilon} - q_1 q_2(\varepsilon_1 - \varepsilon_2)^2/(3\bar{\varepsilon}), \tag{3}$$

which shows that $\varepsilon \le \bar{\varepsilon}$ if $\varepsilon_1 \ne \varepsilon_2$. Note that this result (equation 2.11 of [6]) was obtained without appeal to any dielectric or electrodynamic model. We use it later to compare various mixture rules.

We consider the use of a liquid of known permittivity ε_1 which may be varied, e.g. by varying the composition of two miscible liquids. This liquid may suspend a powder of grains of permittivity ε_2. If the permittivity of the suspension is denoted by ε, then in the case of a perfect match between the liquid and the powder permittivity we have

$$\varepsilon = \varepsilon_1 = \varepsilon_2 . \tag{4}$$

In the case of imperfect match, $\varepsilon_1 \ne \varepsilon_2$, and a Taylor expansion of ε around ε_2 gives

$$\varepsilon = \varepsilon_2 + \sum_{m=1}^{\infty} c_m(\varepsilon_1 - \varepsilon_2)^m \tag{5}$$

in which $c_1 = 1$, and the other coefficients are given by

$$c_m = \frac{(-1)^{m+1}}{m!}\left(\frac{\mathrm{d}^m\varepsilon}{\mathrm{d}\varepsilon_1^m}\right)_{\varepsilon_2} . \tag{6}$$

Equation (5) reiterates that $\varepsilon = \varepsilon_2$ for $\varepsilon_1 = \varepsilon_2$. In practice it is necessary that the coefficients c_m must differ significantly from zero in order that sufficient permittivity mismatch occurs (i.e. $\varepsilon_2 > \varepsilon_1$ or $\varepsilon_2 < \varepsilon_1$) to identify the experimental permittivity match.

2.2. *Existing dielectric mixture rules*

To describe the mixture permittivity $\varepsilon(\varepsilon_1,\varepsilon_2,q_1,q_2)$ in terms of the component permittivities ε_1, ε_2, and volume fractions q_1 and q_2 a large number of mixture rules exist [7]. The various approaches are effective medium approximations in which the heterogeneous length-scale is much smaller than the wavelength of the applied electric field. These rules give approximate c_m to validate the experimental permittivity mismatch measurement method.

Some rules may be written [8]

$$q_2 = \frac{(\varepsilon-\varepsilon_1)[\varepsilon_2 + 2\varepsilon_1 + \nu(\varepsilon-\varepsilon_1)]}{(\varepsilon_2-\varepsilon_1)[\varepsilon + 2\varepsilon_1 + \nu(\varepsilon-\varepsilon_1)]} \tag{7}$$

in which $\nu = 0$, 1, 2, or 3 according to the internal electric field assumed in their formulation. For $\nu = 0$ the Maxwell Garnett rule is obtained. Suffix #1 refers to an external component, while #2 refers to the included component. Other relations have been proposed [7] such as

$$q_2 = (\varepsilon^n - \varepsilon_1^n)/(\varepsilon_2^n - \varepsilon_1^n) \tag{8}$$

and the Lichtenecker rule

$$q_2 = \log(\varepsilon/\varepsilon_1)/\log(\varepsilon_2/\varepsilon_1) . \tag{9}$$

Equation (8) if $n = ½$ becomes the complex refractive index rule used in electromagnetic logging formation-evaluation [9,10] and if $n = ⅓$ it becomes the Looyenga rule [11] which has a theoretical basis [12]. These and similar rules have been summarized [7]. Equations (8) and (9) are symmetric in the disposition of the components #1 and #2, i.e. the value of ε remains the same if the suffices are exchanged, e.g. from 1 to 2 or *vice versa*. We note that the relation of (7) is generally non-symmetric, as expected from a rule that distinguishes between the internal and the external components. The exception is for $\nu = 2$, when the symmetric Polder-van Santen or symmetric Bruggeman rule is

$$q_1 \frac{(\varepsilon_1 - \varepsilon)}{(\varepsilon_1 + 2\varepsilon)} + q_2 \frac{(\varepsilon_2 - \varepsilon)}{(\varepsilon_2 + 2\varepsilon)} = 0. \tag{10}$$

The various dielectric functions $\varepsilon(\varepsilon_1,\varepsilon_2,q)$ possess upper and lower bounds. Statistical continuum theory [5, 6] shows the bounding functions depend on the volume fractions of the components, and on various correlation functions that describe the geometrical disposition or structure of the dielectric's components. If there is no correlation function information and the only relevant parameters are the volume fractions, then the closest possible bounds for a two-component medium are those of Hashin and Shtrikman [5,6]. If $\varepsilon_2 > \varepsilon_1$ we have the lower (LHS) and upper (UHS) bound H-S mixture rules

$$q_2 = \frac{3\varepsilon_2(\varepsilon - \varepsilon_1)}{(\varepsilon_2 - \varepsilon_1)(\varepsilon + 2\varepsilon_2)}, \tag{11}$$

$$q_1 = \frac{3\varepsilon_1(\varepsilon_2 - \varepsilon)}{(\varepsilon_2 - \varepsilon_1)(\varepsilon + 2\varepsilon_1)}, \tag{12}$$

respectively, and vice-versa for $\varepsilon_2 < \varepsilon_1$. Equation (11) corresponds to the Maxwell Garnett (MAG) rule from (7) with $\nu = 0$ where #1 denotes the external component, while (12) also corresponds to (7) with $\nu = 0$ but where #2 denotes the external component. All the rules above have the same term c_1 given by (6) as

$$c_1 = q_1. \tag{13}$$

From (6) the dielectric permittivity ε sufficiently close ($\varepsilon_1 \cong \varepsilon_2$) to the match condition $\varepsilon_1 = \varepsilon_2$ is

$$\varepsilon = \varepsilon_2 + q_1(\varepsilon_1 - \varepsilon_2) = q_1\varepsilon_1 + q_2\varepsilon_2 = \bar{\varepsilon}, \tag{14}$$

hence regaining the leading term of (3). However, the various rules give different relations for the terms c_2, c_3, and for higher-order terms. Table 1 shows that c_2 obtained from the statistical relation (3) agrees with c_2 from the relations of (11), (12) and with the Looyenga rule with $n = ⅓$.

Table 1. Coefficients of (5) for various mixture rules.

Mixture rule equation	c_2	c_3
3	$-q_1q_2/(3\bar{\varepsilon})$	undefined
8 ($n = ½$)	$-q_1q_2/(4\bar{\varepsilon})$	$-q_1q_2/(8\bar{\varepsilon}^2)$
8 ($n = ⅓$)	$-q_1q_2/(3\bar{\varepsilon})$	$-q_1q_2(9+q_2)/(54\bar{\varepsilon}^2)$
9	$-q_1q_2/(2\bar{\varepsilon})$	$-q_1^3/(6\bar{\varepsilon}^2)$
11	$-q_1q_2/(3\bar{\varepsilon})$	$-q_1q_2(2+q_2)/(9\bar{\varepsilon}^2)$
12	$-q_1q_2/(3\bar{\varepsilon})$	$+q_1q_2^2/(9\bar{\varepsilon}^2)$

3. Permittivity-matching measurements

Electroporation cells (Sigma Aldrich) were used to hold about 3 ml of suspension sample. The impedance was measured of about 0.4 ml of sample between two parallel aluminium electrodes 2 mm apart in the cell, connected to the component fixture 1J1011 of a Wayne Kerr 6500B impedance analyser. Permittivity measurements were made in the frequency range 1 kHz to 50 MHz. Mixtures of various concentrations of n-decane and butan-1-ol were used to obtain a liquid permittivity range of 2–16. The permittivities of the liquids (ε_{liq}) and the mineral suspensions (ε_{sus}) showed negligible frequency dependences above 10 kHz and data were averaged over 0.1–10 MHz.

The matching method was used to study number of nominally pure minerals of grain sizes 1 - 10 μm, the mineral permittivity being denoted by ε_{min} ($\equiv\varepsilon_2$). For example, figure 1 (solid line) for SiO_2 is a plot of $\varepsilon_{sus}-\varepsilon_{liq}$ versus ε_{liq}. The line shows a linear fit to data giving $\varepsilon_{min} = 4.44$ when $\varepsilon_{sus}-\varepsilon_{liq} = 0$. This compares closely with $\varepsilon_{min} = 4.47$, being the mean of $\varepsilon_{11} = \varepsilon_{22} = 4.41$ and $\varepsilon_{33} = 4.60$ for this anisotropic mineral [13]. Similarly good agreement was found between the matching method results and literature data for $CaCO_3$, Al_2O_3, and ZrO_2.

Permittivity-matching measurements were made on mixtures of minerals to simulate real rocks which usually comprise a range of different mineral grains. Figure 1 (broken line) is a plot of experimental results for $\varepsilon_{sus}-\varepsilon_{liq}$ versus ε_{liq} for a mineral mixture comprising 25.2% v/v $CaCO_3$ + 74.8% v/v Al_2O_3. The line shows a linear fit to data giving $\varepsilon_{min} = 9.61$ when $\varepsilon_{sus}-\varepsilon_{liq} = 0$. Literature data [13] gives $\varepsilon_{min} = 8.58$ for $CaCO_3$ and $\varepsilon_{min} = 10.03$ for Al_2O_3, and shows that $\varepsilon_{min} = 9.61$ measured for this mineral mixture lies between ε_{min} of the constituent minerals as expected. The measured ε_{min} compares closely with $\varepsilon_{min} = 9.68$ calculated from (8) with $n = ½$ using the above constituent minerals data. The permittivity for various mixtures of $CaCO_3$, Al_2O_3 and SiO_2 was also measured by the matching method. Figure 2 shows good agreement between measured data for the mixtures and the permittivity predicted by (8) with $n = ½$, using the ε_{min} data above for $CaCO_3$, Al_2O_3 and SiO_2.

The effect of imaginary permittivity on the method can be estimated from mixture rules generalized for a complex permittivity, and show that a loss tangent < 0.1 of either the mineral or the liquid gives $< 0.5\%$ error in the real permittivity measured by the matching method.

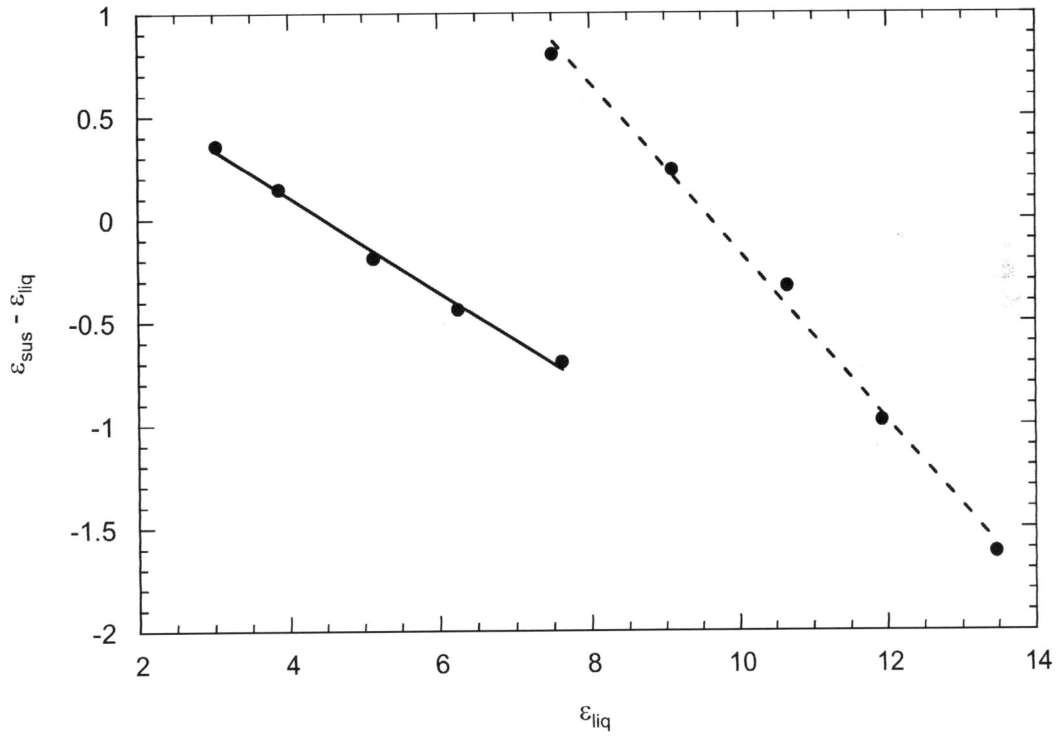

Figure 1. Plots of $\varepsilon_{sus}-\varepsilon_{liq}$ versus ε_{liq} for SiO_2 powder (solid line) and for a powder mixture of 25.2% $CaCO_3$ + 74.8% Al_2O_3 (broken line).

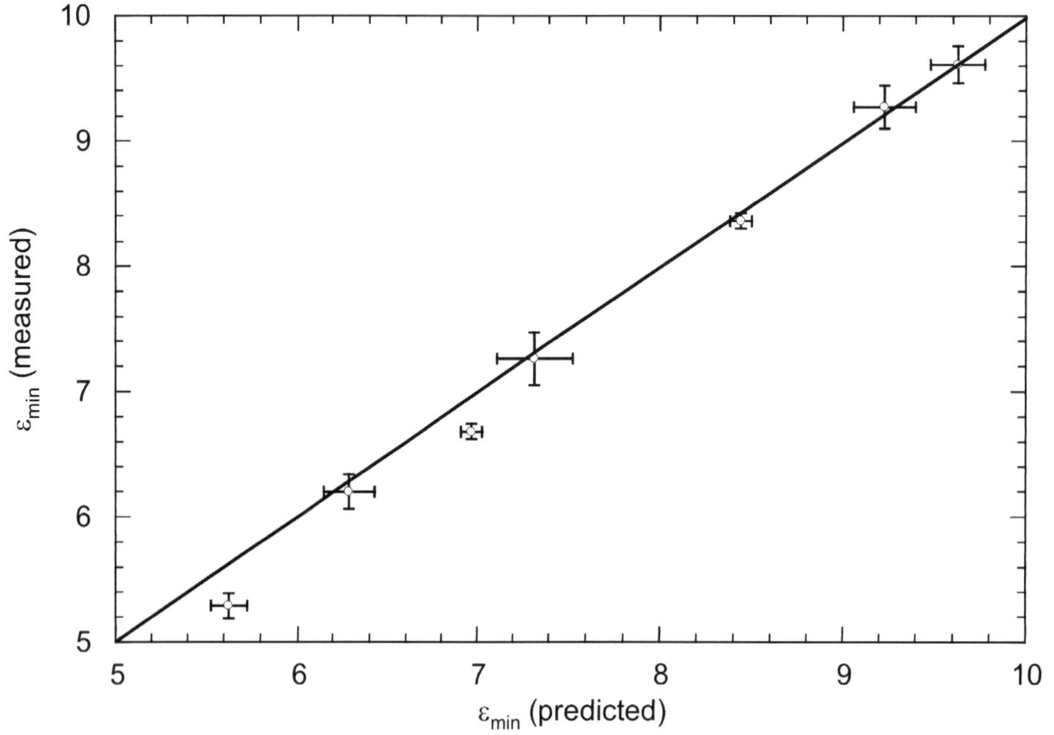

Figure 2. Plot of data markers with error bars to compare measured with predicted data for permittivities of two-part powder mixtures. The trend line is of unit gradient.

4. Summary

A permittivity-matching method is described to obtain the permittivity ε_{min} of mineral particles. The method is independent of particle shape, size, and irregularity. It measures the permittivity ε_{sus} of suspensions of particles for suspending liquids of different permittivity ε_{liq} and determines when $\varepsilon_{\text{liq}} = \varepsilon_{\text{sus}} = \varepsilon_{\text{min}}$. It is supported by dielectric statistical continuum theory and does not assume the validity of any particular dielectric model (or mixture rule). The method was investigated and validated for various mineral types and mixtures.

5. References

[1] Archie, G E 1942 *Petroleum Transactions of AIME* **146** 54-62
[2] Hizem M, Budan H, Devillé B, Faivre O, Mossé L, Simon M 2008 *SPE116130*
[3] Böttcher C F J 1952 *Theory of Electric Polarization* (Amsterdam: Elsevier)
[4] Robinson D A, Friedman S P 2003 *J Geophys Res* **108(B2)** 2076 doi:10.1029/2001JB000691
[5] Beran M J 1968 *Statistical Continuum Theories* (New York: Interscience)
[6] Batchelor G K 1974 *Ann Rev Fluid Mech* **6** 227-255
[7] Jylhä L, Sihvola A 2007 *J Phys D: Appl Phys* **40** 4966-4973
[8] Kärkkäinen K, Sihvola A, Nikoskinen K 2001 *IEEE Trans Geosci Remote Sensing* **39** 1013-1018
[9] Birchak J R, Gardner C G, Hipp J E, Victor J M 1974 *Proc IEEE* **62** 93-98
[10] Freedman R, Vogiatzis JP 1979 *Geophys* **44** 969-986
[11] Looyenga H 1965 *Physica* **31** 401-406
[12] Landau L D, Lifshitz E M 1960 *Electrodynamics of Continuous Media* (Oxford: Pergamon)
[13] Kaye G W C, Laby T H 1995 *Tables of Physical and Chemical Constants* (Teddington: NPL)

New opportunities for secure communication networks using shaped femtosecond laser pulses inducing filamentation processes in the atmosphere

H M Alyami V M Becerra and S Hadjiloucas
School of Systems Engineering, The University of Reading, RG6 6AY, UK

E-mail: s.hadjiloucas@reading.ac.uk

Abstract. The current study discusses new opportunities for secure ground to satellite communications using shaped femtosecond pulses that induce spatial hole burning in the atmosphere for efficient communications with data encoded within super-continua generated by femtosecond pulses. Refractive index variation across the different layers in the atmosphere may be modelled using assumptions that the upper strata of the atmosphere and troposphere behaving as layered composite amorphous dielectric networks composed of resistors and capacitors with different time constants across each layer. Input-output expressions of the dynamics of the networks in the frequency domain provide the transmission characteristics of the propagation medium. Femtosecond pulse shaping may be used to optimize the pulse phase-front and spectral composition across the different layers in the atmosphere. A generic procedure based on evolutionary algorithms to perform the pulse shaping is proposed. In contrast to alternative procedures that would require *ab initio* modelling and calculations of the propagation constant for the pulse through the atmosphere, the proposed approach is adaptive, compensating for refractive index variations along the column of air between the transmitter and receiver.

1. Introduction

Communication networks are nowadays deployed across different media adopting a variety of propagation modalities e.g., wire, optical fibres, microwave or THz links, etc. Recent advances in ultrafast pulse, frequency comb timing standards, and satellite quantum communications technologies may soon enable geostationary satellites to be used as high bandwidth secure communication links. These are paving the way for new global communication modalities beyond the physical confines of the current fibre optics highways. In systematic studies performed by the fibre optics communications laboratories, there has been evidence of a host of linear and no-linear phenomena taking place simultaneously during propagation such as Rayleigh, Mie and Raman scattering, optical Kerr effect, soliton transmission and optical damage (spatial hole burning) even at modest optical powers [1]. The fibre optics communications community has also considered implications of shortening the pulses to the limit of dispersion within a fibre optic network, providing guidelines for the degree of chirping in the transmitted pulse [2], providing also design rules for super-continuum sources for WDM applications [3,4] and calculation of optical phase jitter in dispersion-managed systems [5].

Advances in generating shaped femtosecond pulses are enabling us to re-consider propagation in the higher atmosphere and troposphere assuming these layers may be modelled over a very large bandwidth as dielectrics. This work looks at the physical processes underlying such spatial hole burning and whether we can direct radiation with low losses by harnessing the properties of the super-continuum generated by these pulses. Our aim is to investigate whether we can harness non-linear phenomena through femtosecond pulse shaping techniques in the strong field regime to establish the opening of an efficient secure communication channel from ground to low earth orbit. This is possible because when a laser beam power exceeds a critical value (10^{14} W/cm^2), it starts to self-focus to a point [6-8]. However, the catastrophic collapse is negated due to the ionization of air creating a plasma which in turn defocuses the beam. A self-induced waveguide in air (filament) occurs when the dynamic balance between self-focusing, plasma defocusing and diffraction is met [6-11].

Different layers in the atmosphere as well as the troposphere can be modelled using 2D or 3D networks of resistors and capacitors [12, 13]. In the linear regime, input/output models derived for particular values of time constants [12, 13] can be used to cast propagation phenomena as inverse problems [14]. Such approach is therefore appropriate once steady state conditions through the propagation medium are established, where small fluctuations in atmospheric density may be compensated. The necessary theory for studying propagation under such conditions through layered media is well established [15-21].

The filamentation process, however is more complex and difficult to predict. Although there has been significant progress in understanding discharge as well as filamentation processes by the dielectrics community [16-17], unfortunately the current state-of-the art imaging methods do not enable us yet to fully reproduce filamentation events through different media in a consistent way so as to use results from propagation models as predictors of filamentation events. An alternative method is therefore proposed in the section below.

Femtosecond pulse shaping algorithm

We propose to use femtosecond pulse shaping techniques using adaptive evolutionary algorithms to optimize spatial hole burning and the propagation of pulses in the atmosphere while also preserving phase delay across spectral intervals. In this approach, evolutionary meta-algorithms for pulse shaping of broadband femtosecond duration laser pulses are utilized to achieve this goal. This is not the first time that femtosecond pulse shaping has been considered to provide closed loop control of a physical process [24-27] although to our knowledge it has never been applied to this type of problem.

The genetic algorithm (GA) searching the evolutionary landscape for desired pulse shapes consists of a population of waveforms (genes), each made from two concatenated vectors, specifying phases and magnitudes respectively over a range of frequencies [28]. Frequency domain operators such as mutation, two-point crossover average crossover, polynomial phase mutation, creep and three point smoothing as well as a time-domain crossover may be combined to produce fitter off-springs at each iteration step [29].

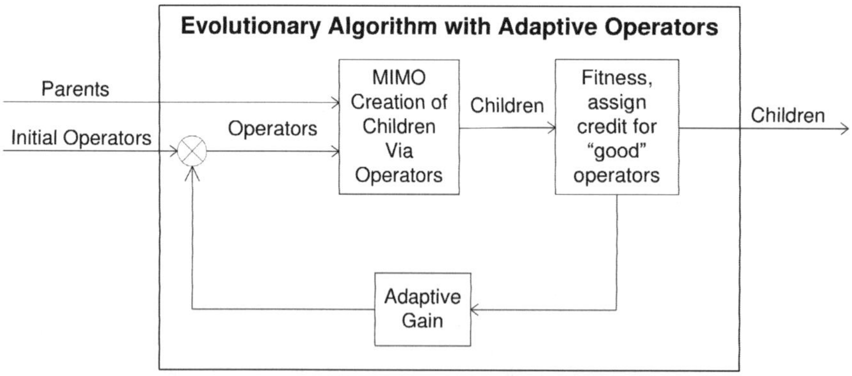

Figure 1. General structure of the proposed evolutionary algorithm for femtosecond pulse shaping where the operators are tuned through an adaptive gain to produce the desirable pulse shape that performs spatial hole burning.

The developed algorithm applies roulette wheel selection, elitists and linear fitness scaling to the gene population associated with the spectral bands associated with the femtosecond pulses. A Differential Evolution (DE) operator that provides a source of directed mutation and wavelet operators are used to speed-up convergence of the pulse to its optimal shape for spatial hole burning. Fast convergence of the algorithm is necessary for maintaining the channel in the presence of changing atmospheric conditions through feedback from the receiving end. Using properly tuned parameters for DE, such meta-algorithm may be used to essentially solve a waveform matching problem in a few generations as shown below.

In the algorithm of Adaptive Operators, any operator that creates a child more fit than the best fit of the previous generation is assigned credit. The credit is then used to update the weights used in choosing the operator that creates future children. While one could theoretically specify an ideal initial set of operators, any static weighting will fail to capture the true dynamic nature of the evolutionary algorithm search space. As the population moves from the initial state to convergence, the ideal set of operators used to create children will change. Adaptive updating not only accounts for the dynamic needs of the search space, but it allows for the inclusion and competition of many different types of operators.

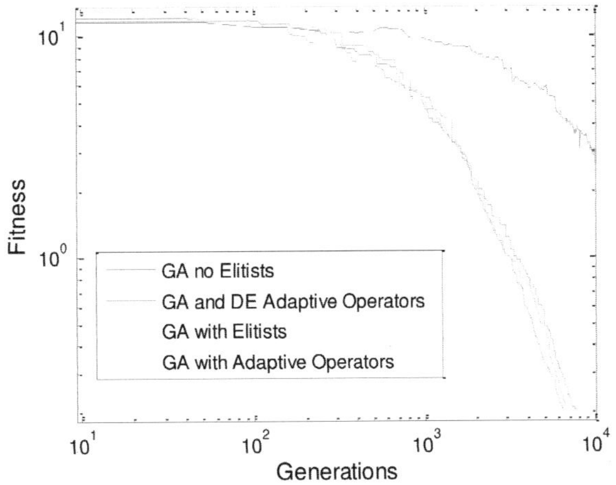

Figure 2. The effect of introducing elitists, differential evolution aadaptive operators to improve the convergence of the algorithm to fewer generators.

Figure 2 shows a graph of four variants of evolutionary algorithms showing their relative speeds in convergence for solving a waveform matching problem. While all methods will converge, the most efficient algorithm is the Genetic Algorithm (GA) hybridized with Differential Evolution (DE), using adaptive operators to adjust the weights (degree of expression) as well as the operators values. Kohonen maps may be used to evaluate the regions where most benefit to the convergent rate can be gained by tuning the individual operators while searching the evolutionary landscape for the optimal pulses. Because differential evolution and the original genetic algorithm outperform each other in different regions of the search space, their combination to a meta-algorithm leads to faster convergence rates in fewer generations.

Hardware implementation

The current algorithm for waveform matching has mainly been implemented in software with the algorithm running within the Matlab environment. By creating a Labview based front end with the capability of running the Matlab script within it, it was possible to perform some preliminary tests in the lab. For example, it was straightforward to perform dispersion compensation of phase delay across a range of frequencies through 4.0 mm of glass using a dazzler (rf-excited acousto-optic transducer) placed in front of a Mira laser system from Coherent generating 100 fs duration pulses. The dazzler parameters were controlled by the output of the evolutionary algorithm through a digital to analog converter. The amorphous glass medium essentially acted as a multiple input-multiple output (MIMO) channel in the same way that a column of air would act on the propagating pulses, introducing different degrees of phase delay at different frequencies. The experiment has shown that the developed algorithm efficiently flattened the spectral phase of the emerging pulse across the 780-820 nm bandwidth as measured with a spectral phase interferometer (a SPIDER from APE Berlin). Information of the phase flatness was used at each iteration step to change the amplitude and phase delay of the individual genes associated with the genetic algorithm.

This experiment partly validated the proposed approach. For the full implementation of the proposed solution, however, an amplifier based femtosecond system would have to be used in conjunction with a shorter duration seed pulse and a photonic crystal fibre to generate the required super-continuum. The repetition rate of the pulses will generally be lower than the standard 100 MHz associated with most commercially available femtosecond systems due to the requirement for an amplification process. A 10 kHz repetition rate femtosecond pulse laser amplifier produces consecutive pulses spatially separated by 30 Km in the atmosphere. An advantage of the proposed solution, however, is that only a few amplitude and phase genes from the entire population would have to be changed in order to compensate for time-varying atmospheric refractive index variations so that convergence of the evolutionary algorithm can be performed within just a few iterations.

Using pulse shaping, it should be possible to use the leading pulse to induce controlled filamentation and spatial hole burning and consecutive pulses to encode information or alternatively use part of the bandwidth in the white light super-continuum to encode and transmit within the same pulse with another part to perform filamentation. Information encoding can be implemented within the super-continuum over a dedicated spectral band using a spatial light modulator placed in a 4*f* optical system [30]. This is achieved by encoding the time-domain pulse in space through a diffraction grating, then changing the orientation of the individual pixels of the liquid crystal mask accordingly to induce attenuation and phase delay in a binary fashion, thus encoding the information in the modulator's spectral bins. This is quite straightforward using a function generator providing the appropriate information to the mask. A second grating would then be used to re-encode this information in the time domain for the transmission.

Alternative implementations of the proposed evolutionary algorithm where the information can be encoded through radio-vorticity [31-32] are also straight-forward to implement using appropriately modulated masks that control orbital angular momentum [33]. Reducing the dynamics using system identification procedures [34] for the MIMO channel associated with the propagation on the basis of the dielectric properties of the atmosphere is also appropriate and relevant to the implementation of the technique as it would reduce the search space of the evolutionary algorithm.

Acknowledgment
H. M. Alyami acknowledges the scholarship support received from the Ministry of Higher Education, Kingdom of Saudi Arabia that made this work possible.

References

[1] R. Kashyap, Nonlinear Optical Fibres, in '*Trends in Optical Fibre Metrology and Standards*,' pp. 69-102, O.D.D. Soares (Ed.), Kluwer Academic Publishers (1995).
[2] S. T. Cundiff, B. C. Collings, L. Boivin, M. C. Nuss, K. Bergman, W. H. Knox, and S. G. Evangelides 'Propagation of Highly Chirped Pulses in Fiber-Optic Communications Systems,' *Journal of Lightwave Technology*, **17**, (5), 811 (1999)
[3] S. Taccheo and L. Boivin Investigation and design rules of supercontinuum sources for WDM applications, Fiber Lasers and Nonlinear Effects (ThA), Optical Fiber Communication Conference, Baltimore, Maryland, March 7, 2000.

[4] S. Hadjiloucas, A Hashem, VM Becerra, and RKH Galvão, Channel equalization for indoor lighting communications networks, *J. Phys.: Conf. Ser.* **450** 012051.

[5] M. Hanna, D. Boivin, P.-A. Lacourt, and J.-P. Goedgebuer, 'Calculation of optical phase jitter in dispersion-managed systems by use of the moment method,' *JOSA B*, **21**, (1), 24-28 (2004).

[6] J. Kasparian M. Rodriguez, G. Méjean, J. Yu, E. Salmon, H. Wille, R. Bourayou, S. Frey, Y.-B. André, A. Mysyrowicz, R. Sauerbrey, J.-P. Wolf, L. Wöste, 'White-Light Filaments for Atmospheric Analysis,' *Science* **301**, 61-64, (2003).

[7] M. Rodriguez, R. Bourayou, G. Méjean, J. Kasparian, J. Yu, E. Salmon, A. Scholz, B. Stecklum, J. Eislöffel, U. Laux, A.P. Hatzes, R. Sauerbrey, L. Wöste and J.-P. Wolf, 'Kilometer-range nonlinear propagation of femtosecond laser pulses', *Phys. Rev. E* **69**, 036607 (2004).

[8] A. Couairon and A. Mysyrowicz, 'Femtosecond filamentation in transparent media, '*Phys. Rep.* **441**, 47-189, (2007).

[9] L. Bergé, S. Skupin, R. Nuter J. Kasparian and J.-P. Wolf, 'Ultrashort filaments of light in in weakly ionized optically transparent media, '*Rep. Prog. Phys.* 70, 1633, (2007).

[10] B. Shim, S. E. Scrauth and A.L. Gaeta, 'Filamentation in air with ultra-short mid-infrared pulses,' *Optics Express*, **19** (10), 9118 -9126 (2011).

[11] S. L. Chin, T.-J. Wang, C. Marceau, J. Wu, J.S. Liu, O. Kosareva, N. Penov, Y.P. Chen, J.-F. Daigle, S. Yuan, A. Azarm, W.W. Liu, T. Seideman, H.P. Zeng, M. Richardson, R. Li, Z.Z. Xu., 'Advanced in intense femtosecond laser filamentation in air,' *Laser Phys.* **22**, 1 (2012).

[12] S. Hadjiloucas, G.C. Walker, J. W. Bowen and R.K.H. Galvão, 'System identification algorithms for the analysis of dielectric responses from broadband spectroscopies', Dielectrics 2011: *Journal of Physics: Conference Series*, **310** (2011) 012002.

[13] R.K.H. Galvão S. Hadjiloucas, K.H. Kienitz, H.M. Paiva and R.J.M. Afonso, Fractional Order Modeling of Large Three-Dimensional RC Networks, IEEE Transactions on Circuits and Systems I **60** (3), 624-637, (2013)

[14] E. Tuncer, 'Distribution of Relaxation Times: An Inverse Problem,' *IEEE Transactions on Dielectrics and Electrical Insulation* **19**, (4) 1221-1225, (2012).

[15] A. Boivin, J. Dow and E Wolf, 'Energy flow in the neighborhood of the focus of a coherent beam,' *J. Opt. Soc. Am.* **57**, 1171-1175 (1967).

[16] E. Wolf and Y. Li, 'Conditions for the validity of the Debye integral representation of focused fields,' *Opt. Commun.*, **39**, 205-210, (1981).

[17] J.J. Stamnes, and V. Dhayalan, 'Focusing of electric dipole waves,' *Pure Appl. Optics*, **5**, 195-226, (1996).

[18] V. Dhayalan and J.J. Stamnes, 'Focusing of mixed dipole waves,' *Pure Appl. Optics*, **6**, 317-345, (1997).

[19] V. Dhayalan and J.J. Stamnes, 'Focusing of electric dipole waves in the Debye and Kirchoff approximations,' *Pure Appl. Optics*, **6**, 347-372, (1997).

[20] G.C. Sherman, J.J. Stamnes, and Lalor 'Assymptotic approximations of angular spectrum representations,' *J. Math. Phys.*, **17**, 760-776, (1976).

[21] T.D. Visser, S.H. Wiersma, 'Diffraction of convergent electromagnetic waves,' *J. Opt. Soc. Am. A*, **9**, 2034-2047, (1991).

[22] S.J. Dodd, "A Deterministic Model for the Growth of Non-conducting Electrical Tree Structures", *J. Phys. D: Appl. Phys.*, **36**, 129–141, (2003).

[24] D. Meshulach, and Y. Silberberg, 'Coherent quantum control of multi-photon transitions by shaped ultra-short optical pulses,' *Phys. Rev. A*, **60**, 1287-1292 (1999).

[25] D. Zeidler, S. Frey, K.L. Kompa and M. Motzkus, 'Evolutionary algorithms and their application to optimal control studies,' *Phys. Rev. A*, **64**, 023420, (2001).

[26] Haverkamp, N. and Telle, H. R. 'Complex intensity modulation transfer function for supercontinuum generation in microstructure fibers,' *Opt. Express*, **12**, 582-587, (2004).

[27] Kim, Y. S. and Rabitz, H. 'Closed Loop Learning Control with Reduced Space Quantum Dynamics,' *J. Chem. Phys.*, **117**, 1024–1030, (2002).

[28] A. Shaver, S. Hadjiloucas, G.C. Walker and J.W. Bowen, 'Femtosecond Pulse Shaping using a Differential Evolution Algorithm and Wavelet Operators,' *Electronics Letters*, **48** (21) 1357-1358 (2012).

[29] B.J. Pearson, J.L. White, T.C. Weinacht, and P.H. Bucksbaum, 'Coherent control using adaptive learning algorithms' Phys. Rev. A, 63, 63-74 (2001).

[30] A.M. Weiner, 'Femtosecond pulse shaping using spatial light modulators,' Rev. Sci. Instrum., 71, 1929-1960 (2000).

[31] M. D'Amico, A. Leva, B.Micheli, "Free-space optics communication systems: First results from a pilot field-trial in the surrounding area of Milan, Italy," *IEEE Microwave and Wireless Components Lett.* **13**, 305 (2003).

[32] M. Tamburini, E. Mari, A. Sponselli, B. Thiedé, A. Bianchini and F. Romanato, 'Encoding many channels on the same frequency through radio vorticity: first experimental test', *New Journal of Physics*, **14**, 033001, (2012).

[33] G. Gibson, J. Courtial, M. J. Padgett, M. Vasnetsov, V. Pas'ko, S. M. Barnett, S. Franke-Arnold, Free-space information transfer using light beams carrying orbital angular momentum, *Opt. Express*, **12**, (22) 5448, (2004).

[34] S. Hadjiloucas, G.C. Walker, J. W. Bowen V.M Becerra A. Zafiropoulos and R.K.H. Galvão, 'High signal to noise ratio THz spectroscopy with ASOPS and signal processing schemes for mapping and controlling molecular and bulk relaxation processes,' Dielectrics 2009: Measurement Analysis and Applications, *Journal of Physics: Conference Series*, **183** (2009) 012003.

[35] R. K. H. Galvão, K. H. Kienitz, S. Hadjiloucas, G.C. Walker, J.W. Bowen, S. F. C. Soares, M. C. U. Araújo, 'Multivariate analysis of random three-dimensional RC networks in the time and frequency domains,' *IEEE Transactions on Dielectrics and Electrical Insulation* **20**, (3), pp. 995-1008, (2013).

Computational study of dielectric function and optical properties of a graphane nano structure containing graphene quantum dot

Jyoti Dhar Sharma*[1], Munish Sharma[1], Naveen Kumar[1] and P K Ahluwalia[2]
[1]Deptt. of Physics, Shoolini University, Bajhol, Distt. Solan, H. P., 173212 (India)
[2]Deptt. of Physics, Himachal Pradesh University, Shimla 171005, (India)

* sharmajyotid@gmail.com

Abstract: Ab-Initio computational study of dielectric function and optical properties of a graphane nano structure containing graphene quantum dot has been undertaken within Density Functional Theory using SIESTA code. Band structure, PDOS, real and imaginary parts of dielectric function, reflectance and energy loss have been calculated and frequencies corresponding to peak positions have been tabulated for each case. A comparison has been made with the corresponding properties of pristine graphene.

Keywords: Quantum Dots, Density Functional Theory, SIESTA, Electronic Density of States, Optical Properties, Dielectric Function.
PACS: 73.21.La 71.15.Mb 31.15E 78.67.Wj 77.22.Ch

1. Introduction

Graphene [1-6], a 2D crystal, is an allotrope of carbon with atoms arranged in a regular hexagonal pattern similar to graphite but in a one-atom thick planar sheet of sp^2-bonded carbon atoms that are densely packed in a honeycomb crystal lattice. It has gained attention due to unique electronic and optical properties [6-9]. Graphene exhibits fascinating properties such as a relativistic massless-dispersion for its charge carriers and ballistic transport on large distances. In addition to numerous experimental and theoretical studies on the physical properties of graphene, efforts have been to synthesize various types of derivatives of graphene such as graphane [6, 10-17]. The first theoretical description of graphane was reported in 2007 [10] and its preparation was reported in 2009 [11]. Graphane is a 2-dimensional polymer of hydrogenated graphene with the formula unit (CH) repeated periodically. Full hydrogenation from both sides of a graphene sheet results in graphane but partial hydrogenation leads to hydrogenated graphene.

The introduction of a cluster of vacancies in hydrogen sub lattice of graphane leads to formation of graphene quantum dots (QD's) [18]. QD's or artificial atoms are one of the most intensely studied systems [19]. QD's have enormous potentials for applications ranging from novel lasers to quantum information processing. Here we explore the electronic and optical response of QD's as vacancies in hydrogen sub lattice of graphane.

In the present work ab initio computational study of nano structures of graphene (50C atoms, 5x5x1 super cell) and graphane containing graphene QD (50C+38H atoms in chair conformation) [figure 1(a-c)] has been undertaken within Density Functional Theory using SIESTA code. QD, a cluster of vacancies in hydrogen sub lattice, was modelled by pulling out hydrogen atoms from graphane. Band structure, PDOS, real and imaginary parts of dielectric function, reflectance and energy loss have been calculated and frequencies corresponding to peak positions have been tabulated for each case. A comparison has been made with the corresponding properties of pristine graphene.

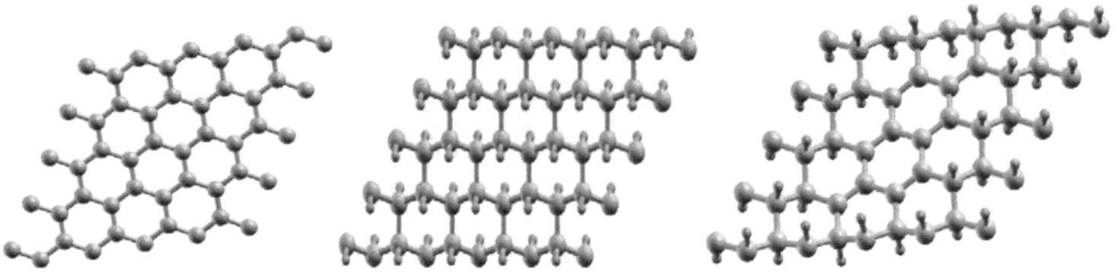

(a) (b) (c)

Figure 1 Optimized structures of **(a)** Pristine Graphene **(b)** Pristine Graphane, 50C+50H atoms **(c)** Graphane containing Quantum Dot, 50C+38H atoms.

2. Computational Details

In this work, we have used Troullier Martin, norm conserving, relativistic pseudopotentials in fully separable Kleinman and Bylander form for Carbon and Hydrogen. Our calculations have been performed in the framework of Density Functional Theory within the generalized gradient approximation (GGA) according to the Perdew, Burke, Ernzerhof (PBE) parameterization using SIESTA code and methods [20, 21]. The graphane sheet is described within the supercell (5x5x1) approach, by a single layer of graphane with 50C atoms. Interaction between adjacent layers is hindered by a large spacing of ~20 Å. A 250 Ry mesh cutoff has been used for the reciprocal space expansion of the total charge density. Brillion zone has been sampled by using 11×11×1 Monkhorst-Pack of k points. Localized atomic orbitals basis set has been used with confinement energy of 0.02 Ry. Minimization of energy is carried out by giving sufficient number of SCF iterations using standard conjugate-gradients technique [20, 21]. Optical calculations have been carried out using 33x33x3 optical mesh and 0.2 eV optical broadening.

3. Results and discussion

3.1 Electronic Band Structure and Interband Transitions

The figure 2 (a & b) shows the plots of band structure and partial electron density of states (PDOS) for pristine graphene [figure 2(a)] and graphane having graphene QD [figure 2(b)]. The band gap decreases with the introduction of a cluster of vacancies in hydrogen sub lattice of graphane.

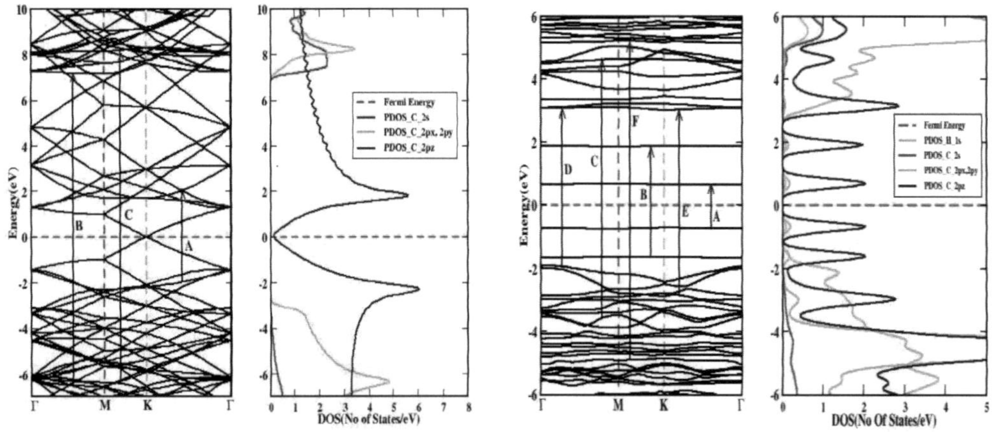

(a) (b)

Figure 2 Band Structure and Electron Density of States for **(a)** Pristine Graphene **(b)** Graphane having QD [Here A, B, C….. : show electronic transitions which correspond to peak positions in the spectrum of optical parameters.]

Calculated multiple sharp peaks near the Fermi energy show the characteristic of quantum dot as an artificial atom. It is clear from the plots of PDOS that the multiple peaks near Fermi energy are due to C_2p$_z$ electrons [figure 2(b)]. The lines A, B & C in figure 2(a) show the electronic transitions which correspond to peak positions in the spectrum of optical parameters at frequencies 4.0 eV, 13.63 eV and 14.49 eV respectively (table 1) for pristine graphene. The peak near 4.0 eV is approximately associated with the transition from valance band near -2.50 eV to conduction band near 1.50 eV while peak at 13.63 eV is associated with the transition from valance band near -6.4 eV to conduction band near 7.23 eV. The peak near 14.49 eV is approximately associated with the transition from valance band near -6.40 eV to conduction band near 8.10 eV.

The figure 2(b) shows the band structure and partial electron density of states (PDOS) for graphane containing graphene QD. The lines A, B, C, D, E & F in figure 2(b) show the electronic transitions corresponding to peak positions in the spectrum of optical parameters at frequencies 1.43 eV, 3.54 eV, 8.10 eV, 4.97 eV, 6.27 eV and 10.09 eV respectively (table 1). The peak near 1.43 eV is approximately associated with the transition from valance band near -0.68 eV to conduction band near 0.63 eV while peak at 3.54 eV is associated with the transition from valance band near -1.59 eV to conduction band near 1.86 eV . The peak near 8.10 eV is approximately associated with the transition from valance band near -3.58 eV to conduction band near 4.62 eV while the peak near 4.97 eV is approximately associated with the transition from valance band near -1.85 eV to conduction band near 3.12 eV. The peak near 6.27 eV is approximately associated with the transition from valance band near -2.81 eV to conduction band near 3.02 eV while the peak near 10.09 eV is approximately associated with the transition from valance band near -4.85 eV to conduction band near 5.20 eV.

3.2 Optical Properties

3.2.1 Real & Imaginary parts of dielectric function

The figure 3(a) shows the plot of real part of dielectric function spectrum for in-plane polarization (E⊥c) and out-of-plane polarization (E || c) for pristine graphene (red curve) and graphane containing graphene QD.

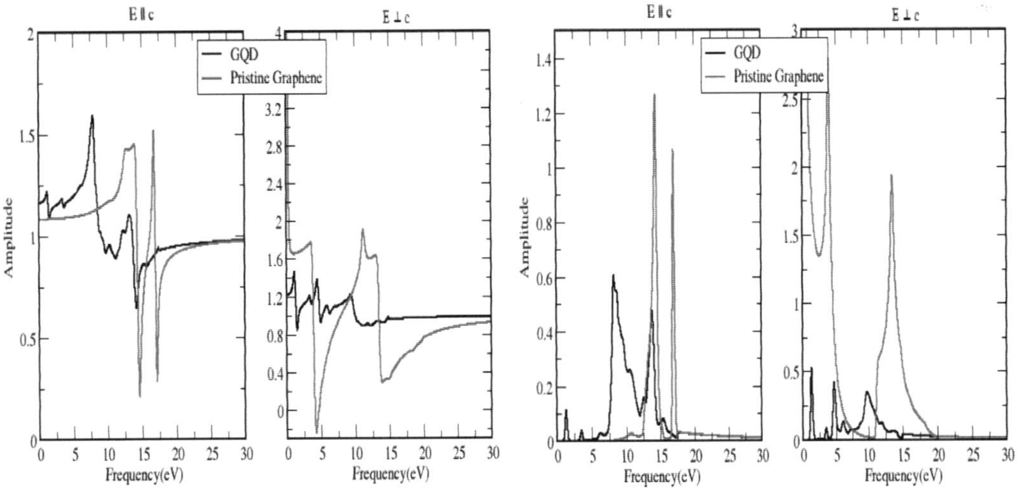

(a) **(b)**

Figure 3 Plot of **(a)** Real part of dielectric function (ε_1), **(b)** Imaginary part of dielectric function for in-plane polarization (E⊥c) and out-of-plane polarization (E || c) for pristine graphene (red curve) and graphane containing graphene QD (black curve).

In the case of pristine graphene, for in-plane polarization (E⊥c), there is a dip in the amplitude of the real part of dielectric function and it attains a small negative value at 4.8 eV indicating that there is plasmonic excitation near 4.8 eV. Also dip in ε_1 at 14.49 eV and 17.01 eV for E || c and 13.63 eV for E⊥c indicate plasmons excitations at these energies.

Corresponding to graphane having graphene QD, the dip in ε_1 at 10.54 eV and 14.04 eV for out-of-plane polarization (E || c) while at 1.47 eV, 4.97 eV and 10.09 eV for in-plane polarization indicate plasmons excitations at these energies.

The figure 3(b) shows the plot of imaginary part of dielectric function (ε_2) for in-plane polarization (E⊥c) and out-of-plane polarization (E || c) for pristine graphene (red curve) and graphane containing graphene QD. Corresponding peak positions are given in table 1. For in-plane polarization, in pristine graphene, peaks appear at 4.0 eV and 13.63 eV, which are in good agreement with recent plane wave calculated results (4.0 eV and 13.8 eV) [7] and experimentally observed values 4.6 eV and 14.6 eV [8]. For (E || c), the spectrum of imaginary part of dielectric function shows peaks at 14.49 eV and 17.01 eV [figure 3(a)] for pristine graphene.

For graphane containing graphene QD, the peaks are observed at frequencies 1.43 eV, 3.54 eV, 8.34 eV & 13.79 eV for out-of-plane polarization while at frequencies 1.39 eV & 4.81 eV for in-plane polarization

3.2.2 Reflectance spectrum & Electron Energy Loss (EEL)

The figure 4(a) shows the plot of reflectance spectrum for in-plane polarization (E⊥c) and out-of-plane polarization (E || c) for pristine graphene (red curve) and graphane containing graphene QD. For out-of-plane polarization in pristine graphene peaks appear at 14.58 eV and 17.29 eV, while for graphane containing graphene QD peaks appear at frequencies 8.1 eV and 13.96 eV. For in-plane polarization in pristine graphene peaks appear at 4.31 eV and 13.80 eV, while for graphane containing graphene QD peaks appear at frequencies 1.39 eV and 4.81 eV.

The figure 4(a) shows the plot of electron energy loss (EEL) spectrum for in-plane polarization (E⊥c) and out-of-plane polarization (E || c) for pristine graphene (red curve) and graphane containing graphene QD.

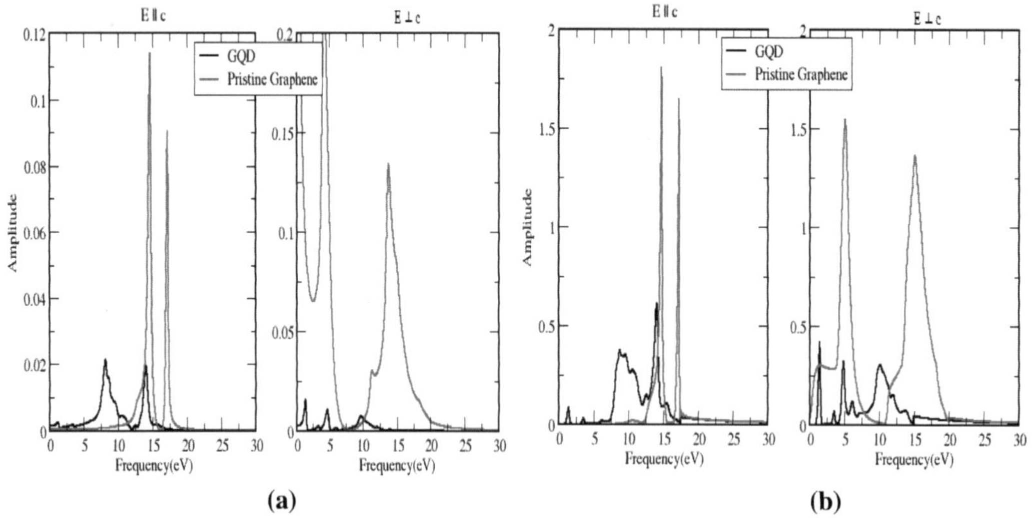

Figure 4 Plot of **(a)** Reflectance spectrum, **(b)** Electron Energy Loss (EEL) Spectrum for in-plane polarization (E⊥c) and out-of-plane polarization (E || c) for pristine graphene (red curve) and graphane containing graphene QD(black curve).

Electron energy loss (EEL) function is proportional to the inverse of dielectric function and corresponds to the collective excitations of electrons of system. In the case of pristine graphene a sharp resonance peak at 5.12 eV and relatively broad peak around 15.16 eV have been found for in-

plane polarization (E⊥c) (see figure 4(b)), which are in very good agreement with recently calculated plane wave results 4.8 eV and 15.0 eV [7] and also with experimentally measured values of 4.7 eV and 14.5 eV respectively [8].
For out-of-plane polarization (*E* || *c*) resonance peaks have been found at 14.72 eV and 17.29 eV, while a plane wave calculation reveals resonance at 11.7 eV and 14.7 eV [7].
For graphane containing graphene QD, the sharp peaks are observed at frequencies 1.43 eV, 8.83 eV & 14.12 eV and a broad peak is observed around 10 eV for out-of-plane polarization. For in-plane polarization sharp peaks are observed at frequencies 1.47 eV, 3.50 eV, 4.97 eV, 6.27 eV & 10.09 eV. Thus new sharp peaks appear in the frequency region of 0 eV to 5 eV for graphane containing graphene QD for in-plane polarization.
Table 1 shows the peak positions in eV for imaginary part of dielectric function, reflectance and electron energy loss in case of out-of-plane polarization (E || c) and in-plane polarization (E⊥c) for pristine graphene and graphane containing graphene quantum dot (GQD).

3.2.3 Red Shift in Peak Frequencies
It is found that there is red shift in plasmonic resonance frequencies in the case of graphane nano structure containing graphene QD as compared to pristine graphene, although the shifts in the peaks are not uniform as can be seen in figure 4(b) and table 1. As compared to pristine graphene, the spectra of dielectric function show red shift of 0.56 eV for out-of-plane polarization (E || c) and 3.74 eV for in-plane polarization (E⊥c). In EEL spectrum the red shift of 0.60 eV has been found for E || c while red shift of 0.15 eV for E⊥c.

Table 1: Peak positions in eV for imaginary part of dielectric function, reflectance and electron energy loss in case of polarization parallel to c and perpendicular to c for pristine graphene and graphane containing graphene quantum dot. (GQD: *Graphane containing graphene quantum dot*)

Optical Properties	**System**	**Peak Positions when E || c (in eV)**	**Peak Positions when E⊥c (in eV)**	
		Our simulation	Our simulation	Others
Imaginary part of dielectric function	Pristine Graphene	14.49, 17.01	4.0, 13.63	4.0, 13.8 [7] 4.6, 14.6 [8]
	GQD	1.43, 3.54, 8.34, 13.79	1.39, 4.81	
Reflectance Spectrum	Pristine Graphene	14.58, 17.29	4.31, 13.80	
	GQD	8.10, 13.96	1.39, 4.81	
Electron Energy Loss Spectrum	Pristine Graphene	14.72, 17.29	5.12, 15.16	4.8, 15.0 [7] 4.7, 14.5 [8]
	GQD	1.43, 8.83, 14.04	1.47, 3.50, 4.97, 6.27, 10.09	

4. Conclusions
In conclusion, we have performed first principle calculations to study the dielectric function and change in optical response of graphane nanostructure containing graphene QD as compared to pristine graphene.

In the band structure band gap appears near the Fermi energy which is found to be 5.23039 eV in pristine graphane (50C+50H) and 1.36846 eV in graphane containing graphene QD. In the plot of PDOS multiple peaks are observed near the Fermi energy which are due to C_2p_z electrons. As compared to pristine graphene, the peak position in the spectra of dielectric function shows red shift of 0.56 eV for out-of-plane polarization (E || c) and 3.74 eV for in-plane polarization (E⊥c). The peak intensities are reduced as compared to pristine graphene. The graphane nano structure having graphene QD is optically more active in the visible region showing additional peaks between 0 eV to 5 eV as compared to pristine graphene.
These electronic and dielectric properties of graphane nanostructure containing graphene QD can be useful for the applications in microelectronics and optoelectronic devices.
.

Acknowledgements
We acknowledge SIESTA team for SIESTA code.

References
[1] Novoselov K S, Geim A K, Morozov S V, Jiang D, Zhang Y, Dubonos S V, Grigorieva I V and Firsov A A 2004 *Science* **306** 666-669
[2] Geim A K and Novoselov K S 2007 *Nature Materials* **6** 183-191
[3] Geim A K and MacDonald A H 2007 *Physics Today* **60** 35-41
[4] Castro Neto A H, Guinea F, Peres N M R, Novoselov K S and Geim A K 2009 Rev. Mod. Phys. **81** 109
[5] Peres N M 2010 Rev. Mod. Phys. **82** 2673
[6] Abergel D S L, Apalkov V, Berashevich J, Ziegler K and Chakraborty Tapash 2010 *Adv.Phys.* **59** 261
[7] Sedelnikova O V, Bulusheva L G and Okotrub V 2011 *J. Chem. Phys.* **134** 24707
[8] Eberlein T, Bangert U, Nair R R, Jones R, Gass M, Bleloch A L, Novoselov K S, Geim A K and Briddon P R 2008 *Phy. Rev. B* **77** 233406
[9] Mohan Brij, Kumar Ashok and Ahluwalia P K 2012 *Physica E* **44** 1670-1674
[10] Sofo J O, Chaudhari A S and Barber G D 2007 *Phys. Rev. B* **75** 153401
[11] Boukhvalov D W, Katsnelson M I and Lichtenelson A I 2008 *Phys. Rev. B* **77** 035427
[12] Elias D C, Nair R R, Mohiuddin T M G, Morozov S V, Blake P, Halsall M P, Ferrari A C, Boukhvalov D W, Katsnelson M I, Geim A K and Novoselov K S 2009 *Science* **323** 610
[13] Flores M Z S, Autreto P A S, Legoas S B and Galvao D S 2009 *Nanotechnology* **20** 465704
[14] Srinivasan C and Saraswathi R 2009 *Current Science* **97** 302
[15] Chandrachud P, Pujari B S, Haldar S, Sanyal B and Kanhere D G 2010 *J. Phys. Condens.Matter* **22** 465502
[16] Boukhvalov D W and Katsnelson M I 2009 *J.Phys. Condens. Matter* **21** 344205
[17] Sharma Jyoti Dhar, Ahluwalia P K and Kumar Naveen 2011 *AIP Conf. Proc.* **1393** 321
[18] Singh A K, Penev E S and Yakobson B I 2010 *ACS Nano* **4** 3510-3514
[19] Chakraborty T 1999 *Quantum dots: a survey of the properties of artificial atoms* (New York: Amsterdam; Elsevier)
[20] Soler J M, Artacho E, Gale J D, Garcia A, Junquera J, Ordejon P and Portal D S 2002 *J. Phys: Condens. Matter* **14** 2745-2779
[21] User's Guide, SIESTA 3.0-beta-15, www.icmab.es/siesta

Wavelet based detection of changes in the composition of RLC networks

H.M. Paiva[1], M.A.Q. Duarte[2], R.K.H. Galvao[3] and S. Hadjiloucas[4]
[1]*Mectron - Odebrecht Organization São José dos Campos, SP, 12227-000, Brazil*
[2]*Department of Mathematics, Universidade Estadual de Mato Grosso do Sul (UEMS) 79540-000, Cassilandia, MS – Brazil*
[3]*Divisão de Engenharia Eletrônica, Instituto Tecnológico de Aeronáutica, São José dos Campos, SP, 12228-900 Brazil.*
[4]*School of Systems Engineering, The University of Reading, RG6 6AY, UK*

E-mails: s.hadjiloucas@reading.ac.uk ; kawakami@ita.br

Abstract. The current work discusses the compositional analysis of spectra that may be related to amorphous materials that lack discernible Lorentzian, Debye or Drude responses. We propose to model such response using a 3-dimensional random RLC network using a descriptor formulation which is converted into an input-output transfer function representation. A wavelet identification study of these networks is performed to infer the composition of the networks. It was concluded that wavelet filter banks enable a parsimonious representation of the dynamics in excited randomly connected RLC networks. Furthermore, chemometric classification using the proposed technique enables the discrimination of dielectric samples with different composition. The methodology is promising for the classification of amorphous dielectrics.

1. Introduction

In broadband dielectric spectroscopy studies of amorphous materials that lack discernible Lorentzian, Debye or Drude responses, we tend to observe a Jonscher-like admittance response of fractional order as a function of frequency [1-17]. We propose to model such response using a 3-dimensional random RLC network, as depicted in Figure 1. Large networks of resistors, inductors and capacitors can be employed as lumped-parameter representations to model the response of composite materials containing conductive, resonant and insulating grains. For fixed R, L, C values, the network response in the time and frequency domains depends on the fraction of each component type. An incidence matrix analysis is used to model the network of branches with nodes consisting of resistive and capacitive elements distributed across several interconnected layers. The present paper is concerned with a screening problem in which deviations from a typical network composition are to be detected. In a real setting, such deviations may arise as the consequence of faults in the production process of a composite material, which should be promptly corrected to avoid the loss of additional batches.

2. Results from simulating 3-dimensional *RLC* networks

In the current study, wavelet identification of these networks is performed to infer the composition of the networks. The analysis is carried out in terms of a wavelet decomposition of the transient response to a standard excitation.

As can be seen in Figures 2 and 3, an increase in the fraction of resistors (f_R) cannot be clearly detected from changes in the amplitude of the current waveforms $i_s(t)$.

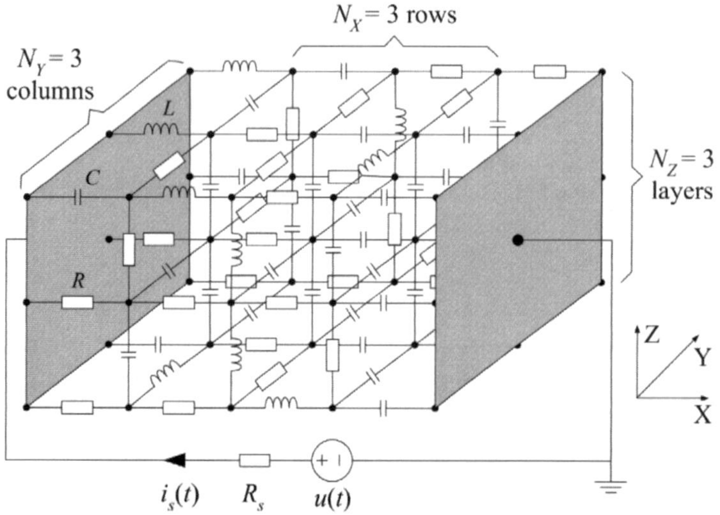

Fig. 1. Example of a 3-dimensional network with random allocation of *R, L, C* elements. The gray plates represent electrodes through which the network is connected to a nominal voltage source with resistance R_S.

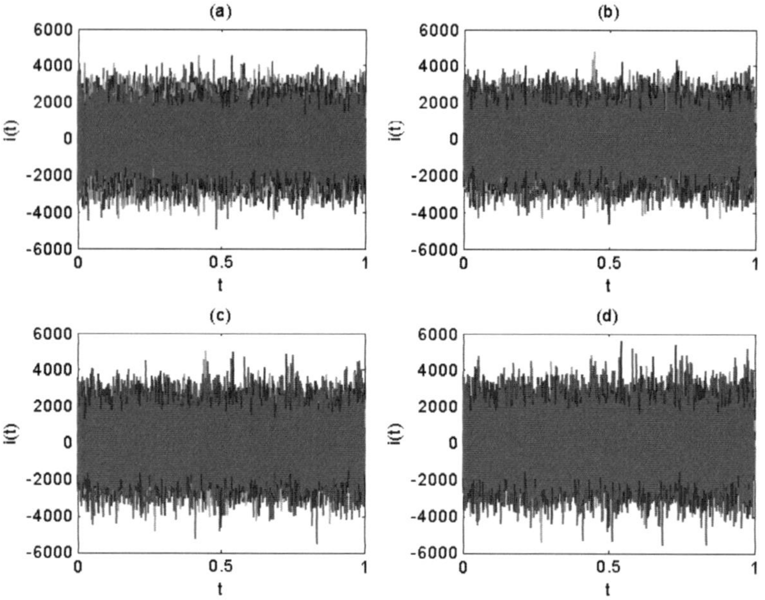

Fig. 2. Current waveforms obtained for 30 realizations of a 3-dimensional network with $N_X = 5$, $N_Y = 5$, $N_Z = 6$ and $R_S = 0.001$, $R = 0.1$, $C = 0.5$, $L = 0.02$ (normalized units) by using a random voltage excitation (white noise). The network composition was set to $f_R = 0.1$, $f_C = 0.45, f_L = 0.45$ in the nominal case (a). In the other cases, the resistor fraction was increased to $f_R =$ 0.2 (b), $f_R = 0.3$ (c) and $f_R = 0.4$ (d), while keeping $f_C = f_L = (1 - f_R)/2$. The sampling frequency was set to 10^4 (normalized unit).

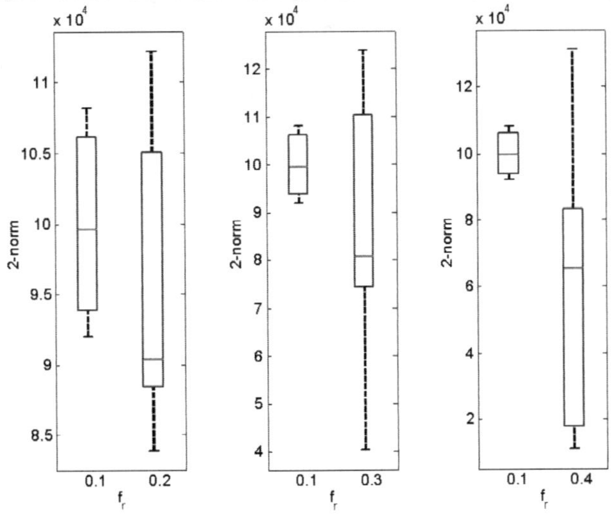

Fig. 3. Boxplots of the 2-norm of the current waveforms for the 30 network realizations. Each frame depicts the boxplot for the networks with nominal composition (left-hand side) and altered composition (right-hand side). On each box, the central red line corresponds to the median, the edges of the blue box indicate the 25th and 75th percentiles, and the black dotted segments extend to the most extreme data values which are not considered to be outliers.

However, by using the wavelet filter bank architecture depicted in Fig. 4 below, it is possible to obtain residues that change in a noticeable manner when f_R is increased as shown in Figures 5a-c.

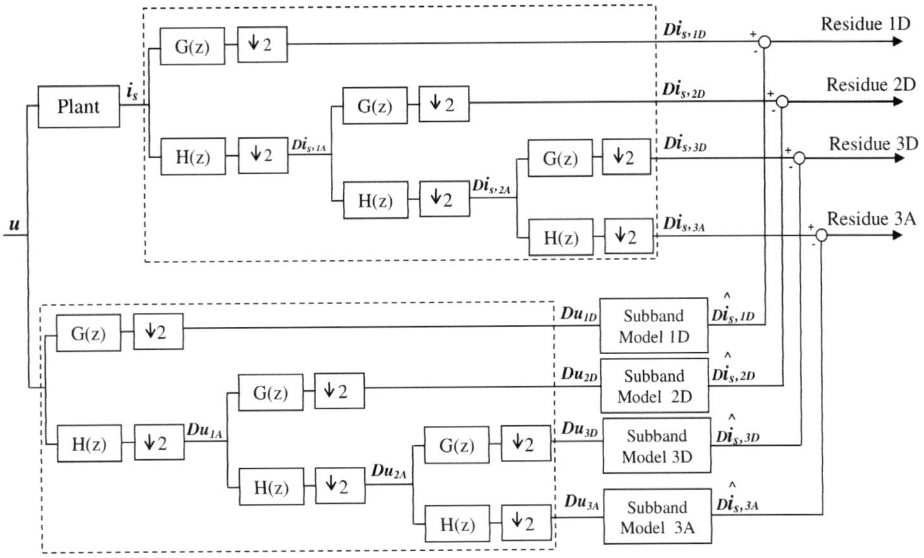

Fig. 4. Wavelet filter bank architecture employed in the generation of residues [18]. The db8 wavelet was employed in all calculations.

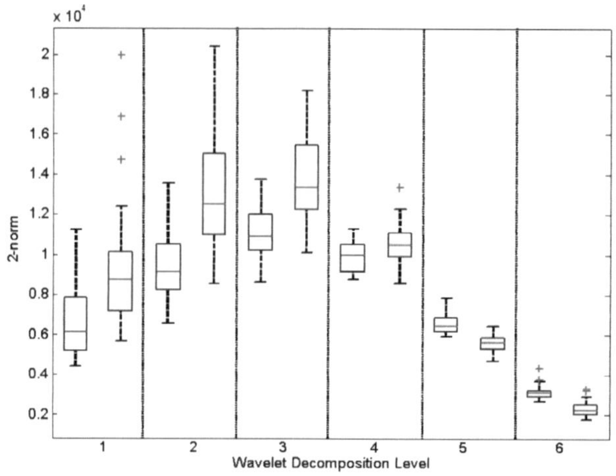

(a)

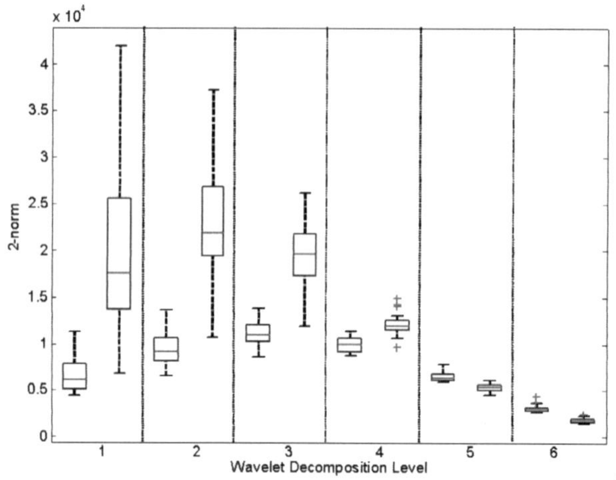

(b)

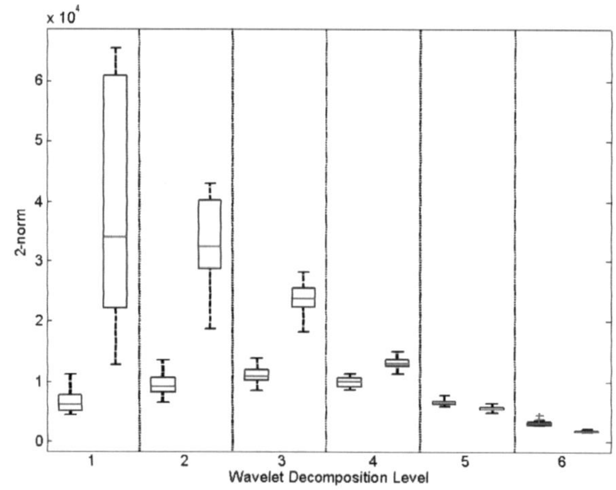

(c)

Fig. 5. Boxplots of the 2-norm of the sub-band residues for the 10 network realizations. Each frame corresponds to a fixed wavelet decomposition level, with a boxplot for the networks with nominal composition on the left-hand side and the networks with altered composition a) ($f_R = 0.2$), b) ($f_R = 0.3$) and c) ($f_R = 0.4$) on the right-hand side. Outliers are indicated with red cross marks.

Wavelet filter banks have been extensively used in signal processing of time domain or frequency domain signals obtained in broadband spectroscopy studies [19-28], so the current application is a direct extension of these works providing an additional link between experimental studies and modelling using *RLC* networks. Further refinement of the technique may be possible by applying feature selection techniques, such as the successive projection algorithms [29-31], to the output of each sub-band of the wavelet models.

3. Conclusion

The results reveal that the proposed optimization procedure is indeed of value to improve sensitivity with respect to changes in the network composition. In a Chemometrics context, the reduced order models enable a simplified representation of the network dynamics, which can be used to infer the composition of the materials. This is of interest from a quality control perspective. The methodology is promising in classification of amorphous dielectrics.

Acknowledgment

The authors acknowledge CNPq for the provision of post-doctoral and research fellowships and FAPESP for the provision of grants 2011/13777-8 and 2011/17610-0 that enabled this work to be performed.

References

[1] Jonscher A K, 1977 "The universal dielectric response", *Nature*, **267**, No. 5613, pp. 673-679.

[2] Lewis T J, 1977 "The dielectric behaviour of non-crystalline solids", *Dielectric and Related Molecular Processes*, **3**, pp. 186-218.

[3] Hurt R L, and Macdonald J R, 1986 "Distributed circuit elements in impedance spectroscopy: a unified treatment of conductive and dielectric systems", *Solid State Ionics*, **20**, 111-124.

[4] Macdonald J R, 2000 "Comparison of the universal dynamic response power-law fitting model for conducting systems with superior alternative models", *Solid State Ionics*, **133**, 79-97.

[5] Tuncer E, Nettelblad B, and Gubanski S M, 2002 "Non-debye dielectric relaxation in binary dielectric mixtures (50-50): Randomness and regularity in mixture topology", *J. Appl. Phys.*, **92**, 4612–4624.

[6] Bouamrane R and Almond D P, 2003 "The emergent scaling phenomenon and the dielectric properties of random resistor-capacitor networks", *J. Phys. Condensed Matter*, **15,** (24), 4089–4100.

[7] Panteny S, Stevens R and Bowen C R, 2005 "The frequency dependent permittivity and AC conductivity of random electrical networks," *Ferroelectrics*, **319**, 199-208.

[8] Almond D P, Bowen C R and Rees D A S, 2006 "Composite dielectrics and conductors: simulation, characterization and design", *J. Phys. D: Appl. Phys.*, **39**, 1295-1304.

[9] Almond D P, Duncan G K, West A R, 1983 "The determination of hopping rates and carrier concentrations in ionic conductors by a new analysis of ac conductivity," *Solid State Ionics*, **8**, (2), 159–164.

[10] Almond D P, Hunter C C, West A R 1984 "The extraction of ionic conductivities and hopping rates from ac conductivity data, *Journal of materials science*," **19**, 3236-3248.

[11] Almond D P, West A R 1983 "Impedance and modulus spectroscopy of "real" dispersive conductors," *Solid State Ionics* **11**, (1), 57–64.

[12] Almond D P, West A R 1983 "Mobile ion concentrations in solid electrolytes from an analysis of ac conductivity," *Solid State Ionics* **9–10**, (1), 277–282,.

[13] Tuncer E and Macdonald J R, 2006 "Comparison of methods for estimating continuous distributions of relaxation times", *J. Appl. Phys.*, **99**, 074106.

[14] McCullen N J, Almond D P, Budd C J, and Hunt G W, 2009 "The robustness of the emergent scaling property of random RC network models of complex materials", *J. Phys. D: Appl. Phys.*, **42**, (6), 64001.1– 64001.8.

[15] Almond D P, Budd C J, Freitag M A, Hunt G W, McCullen N J, Smith N D, 2012 "The Origin of Power-Law Emergent Scaling in Large Binary Networks," *arXiv:1204.5601v1* [cond-mat.dis-nn]

[16] Galvão R K H, Kienitz K H, Hadjiloucas S, Walker G C, Bowen J W, Soares S F C, Araújo M C U, 2013 "Multivariate analysis of random three-dimensional RC networks in the time and frequency domains," *IEEE Trans. Dielectr. Electr. Insul.*, **20**, (3), 995-1008.

[17] Galvão R K H, Hadjiloucas S, Kienitz K H, Paiva H M and Afonso R J M, 2013 "Fractional Order Modeling of Large Three-Dimensional RC Networks," *IEEE Transactions on Circuits and Systems I*, **60** (3), 624-637.

[18] Paiva H M, Galvão R K H, and Rodrigues L, 2009 "A Wavelet-Based Multivariable Approach for Fault Detection in Dynamic Systems", *Controle & Automação*, **20**, (4), 455-464.

[19] Zaengl W S, 2003 "Dielectric Spectroscopy in Time and Frequency Domain for HV Power Equipment, Part I: Theoretical Considerations", *IEEE Electr. Insul. Mag.*, **19**, (5), 5-19.

[20] Galvão R K H Hadjiloucas S, Bowen J W and Coelho C J, 2003 'Optimal discrimination and classification of THz spectra in the wavelet domain,' *Optics Express*, **11**, 1462-1473.

[21] Hadjiloucas S, Galvão R K H, Bowen J W, Martini R, Brucherseifer M, Pellemans H P M, Haring Bolivar P , Kurz H, Digby J, Parkhurst G M, and Chamberlain J M 2003 'Measurement of propagation constant in waveguides using wideband coherent THz spectroscopy,' *Journal of the Optical Society of America B* **20**, 391-401.

[22] Hadjiloucas S, Galvão R K H, Becerra V M, Bowen J W, Martini R, Brucherseifer M, Pellemans H P M, Haring Bolivar P, Kurz H, and Chamberlain J M, 2004 'Comparison of state space and ARX models of a waveguide's THz transient response after optimal wavelet filtering,' *IEEE Transactions on Microwave Theory and Techniques MTT*, **52**, (10), 2409-2419.

[23] Galvão R K H, Hadjiloucas S, Becerra V M and Bowen, J W 2005 'Subspace system identification framework for the analysis of multimoded propagation of THz-transient signals,' *Measurement Science and Technolgy*, **16**, 1037-1053.

[24] Hadjiloucas S, Walker G C, Bowen J W Becerra V M Zafiropoulos A and Galvão R K H, 2009 'High signal to noise ratio THz spectroscopy with ASOPS and signal processing schemes for mapping and controlling molecular and bulk relaxation processes,'*Journal of Physics: Conference Series*, **183** 012003.

[25] Evagorou D, Kyprianou A, Lewin P L, Stavrou A, Efthymiou V, Metaxas A C, and Georghiou G E, 2010 "Feature extraction of partial discharge signals using the wavelet packet transform and classification with a probabilistic neural network", *IET Sci., Meas. Techn.*, **4**, (3), 177-192.

[26] Hadjiloucas S, Walker G C, Bowen J W and Galvão R K H, 2011 'System identification algorithms for the analysis of dielectric responses from broadband spectroscopies', *Journal of Physics: Conference Series*, **310** 012002.

[27] Hao L, Lewin P L, Hunter J A, Swaffield D J, Contin A, Walton C and Michel M, 2011 "Discrimination of Multiple PD Sources Using Wavelet Decomposition and Principal Component Analysis", *IEEE Trans. Dielectr. Electr. Insul.*, **18**, (5), 1702-1711.

[28] Li J, Cheng C, Jiang T and Grzybowski S, 2012 "Wavelet De-noising of Partial Discharge Signals Based on Genetic Adaptive Threshold Estimation", *IEEE Trans. Dielectr. Electr. Insul.*, **19**, (2), 543-549.

[29] Soares S F C, Gomes A A, Galvão Filho A R, Araújo M C U and Galvão R K H, 2013 "The successive projections algorithm", *Trends in Analytical Chemistry*, **42**, 84-98.

[30] Galvão R K H, Araújo M C U, Fragoso W D, Silva E C, José G E, Soares S F C and Paiva H M, 2008 "A variable elimination method to improve the parsimony of MLR models using the successive projections algorithm", *Chemom. Intell. Lab. Syst.*, **92**, 83-91.

[31] Paiva H M, Soares S F C, Galvão R K H and Araújo M C U, 2012 "A graphical user interface for variable selection employing the Successive Projections Algorithm", *Chemom. Intell. Lab. Syst.*, **118**, 260-266.

AUTHOR INDEX

Institute of Physics
Dirac House, Temple Back
Bristol BS1 6BE UK

CURRAN ASSOCIATES INC
proceedings.com